Sheet Metal Hand Processes

Sheet Metal Hand Processes

CLAUDE J. ZINNGRABE • FRED W. SCHUMACHER

10 9 8 7

LIBRARY OF CONGRESS CATALOG CARD NUMBER: 73-2159
ISBN: 0-8273-0220-7

Printed in the United States of America
Published simultaneously in Canada
by Nelson Canada,
A division of The Thomson Corporation

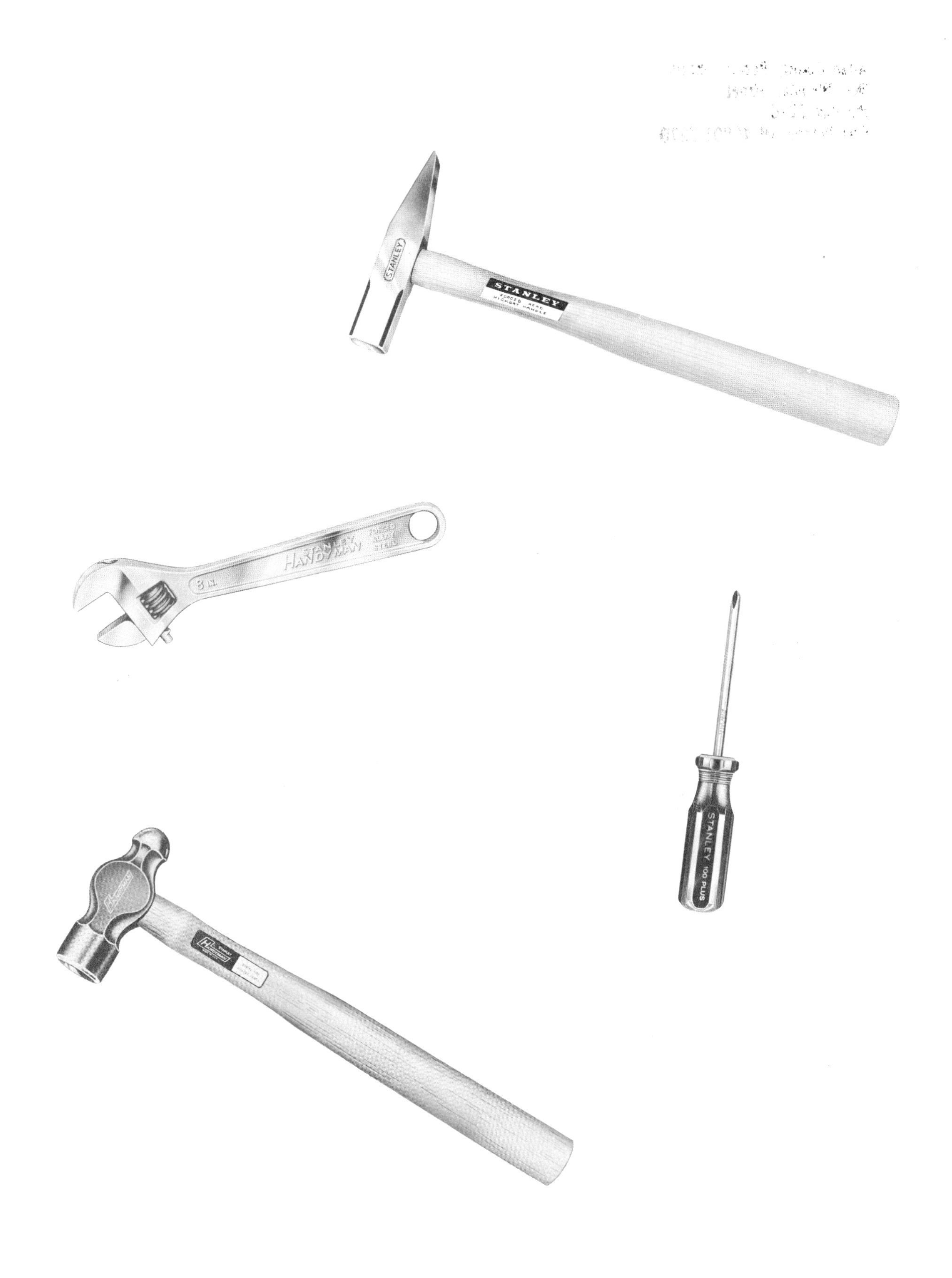

DELMAR PUBLISHERS INC.

PREFACE

A knowledge of hand tools and the ability to use them properly are the fundamentals required of a skilled sheet metal technician. *Sheet Metal Hand Processes* introduces the beginning sheet metal technician to basic instruction in the properties and selection of metal and the processes performed with measuring and marking, bench, cutting, piercing, joining, and soldering tools. Since each hand tool is designed for a specific purpose, the text emphasizes the selection and use of the proper tool for a particular operation.

Sheet Metal Hand Processes, which replaces an earlier text of the same title, has been completely re-written to provide up-dated and expanded instructional material. Vocational sheet metal courses, industrial arts programs with sheet metal units, and apprentice programs will find this text extremely suitable for their use.

The following summarizes the major features of *Sheet Metal Hand Processes:*

- performance objectives are listed at the beginning of each unit to indicate the desired student achievements upon completing the unit
- an excellent selection of illustrations, which are closely correlated to the written material, not only show the various hand tools but also demonstrate the proper use of each tool
- sheet metal processes are described in a step-by-step fashion that is clear, concise, and easily understood
- descriptions of hand tools and their uses include modern tools as well as those that have been used for many years in the sheet metal industry
- safety precautions are emphasized in the use of each hand tool
- summary review questions are provided at the end of each unit; answers to the review questions are provided in the Instructor's Guide.
- an identification review is included in the appendix to evaluate the student's ability to identify the various hand tools and the seams and edges used in sheet metal operations

Sheet Metal Hand Processes was prepared by Claude T. Zinngrabe and Fred Schumacher. Mr. Zinngrabe is a graduate mechanical engineer with thirty-five years of teaching experience in the metal trades at the Washburne Trade School in Chicago, Illinois. Mr. Schumacher is a member of Sheet Metal Workers Local No. 73 with twenty-three years of journeyman field experience and sixteen years of apprentice teaching experience at the Washburne Trade School; presently, he is the chairman of the Sheet Metal Apprentice Department.

Other texts in the Delmar Sheet Metal Technology series include:

Measurement and Layout
Sheet Metal Blueprint Reading for the Building Trades
Mathematics for Sheet Metal Fabrication
Practical Problems in Math for Sheet Metal Technicians
Basic Sheet Metal Skills
Advanced Sheet Metal Skills
Safety for Sheet Metal Workers
Sheet Metal Machine Processes

CONTENTS

The author and editorial staff at Delmar Publishers Inc. are interested in continually improving the quality of this instructional material. The reader is invited to submit constructive criticism and questions. Responses will be reviewed jointly by the author and source editor. Send comments to:

Delmar Publishers Inc.
3 Columbia Circle
Box 15-015
Albany, New York 12212

A sheet metal processing plant. (Courtesy of Inland Steel Company)

UNIT 1 METAL STOCK

OBJECTIVES

After studying this unit, the student should be able to

- List the types and uses of sheet metal stock.
- Describe the advantages of each metal.
- Select the proper metal for the job.

Although it is essential for the sheet metal worker to know the tools and machines of his trade, he should also have a general knowledge of the materials and supplies he must use. It is important to use the correct material for a given job. Usually the job sheet, print, or specifications show the materials to be used. If there is any doubt, the steel mill or supply house will recommend the correct materials to satisfy specified conditions.

SHEET MATERIALS

In sheet metal work 90 percent of all jobs are fabricated from steel, a commercial iron containing carbon in any amount up to 1.7 percent, which is rolled into sheets. Other types of stock used by the sheet metal worker are zinc, copper, aluminum, tin, black iron, and stainless steel.

Sheet materials used in the sheet metal trade are classified as coated sheets and solid sheets. Coated sheets have a base metal, such as steel, with a surface coating of a different metal. This coating protects the base metal from *corrosion*, a gradual eating away of metal by chemical action. Galvanized iron and tin plate are coated sheets. Solid sheets are made of the same material throughout. Examples of solid sheets are copper, aluminum, and stainless steel.

Solid sheets can be cut, formed, sanded, welded, or filed without damaging the corrosion resistance of the sheet. The coated sheet, on the other hand, loses its corrosion resistance if the surface is damaged in any way. This difference is very important in sheet metal fabrication.

Metal sheets may be plain, ribbed, corrugated, or expanded. They range in thickness from .0134 to .1681 inches. The upper limit of stock used in the sheet metal trade is .1382 inches, or 10 gage.

Black Iron

Black iron, steel which contains less than .1 percent carbon, is not extensively used in the sheet metal shop today. Because black iron sheets corrode quickly, they are used for objects that require painting, such as benches, cabinets, and lockers. Although black iron is difficult to solder, it can be welded with considerable ease. This weldability makes it suitable for commercial exhaust systems handling dry waste.

The standard sizes of black iron sheets in inches are 24, 26, 28, 30, 36, 42, and 48 wide by 96 and 120 long. They are available in bundles weighing approximately 150 pounds per bundle. Gages, weights, and thicknesses are found in table 1-1.

Figure 1-1 Ribbed galvanized sheets being cut at the manufacturing plant
(Courtesy of Inland Steel Co.)

Figure 1-2 Stainless steel sheets
(Courtesy of Joseph T. Ryerson & Son, Inc.)

Gage U.S.	Black Iron Thick In.	Black Iron Lb. per Sq. Ft	Galv. Iron Lb. per Sq. Ft.	Gage U.S.	Black Iron Thick In.	Black Iron Lb. per Sq. Ft.	Galv. Iron Lb. per Sq. Ft.
32	.0097	.40625	.56250	19	.0418	1.75	1.90625
31	.0105	.43750	.59375	18	.0478	2.	2.15625
30	.0120	.5	.65625	17	.0538	2.25	2.40625
29	.0135	.5625	.71875	16	.0598	2.5	2.65625
28	.0149	.625	.78125	15	.0673	2.8125	2.96875
27	.0164	.6875	.84375	14	.0747	3.125	3.28125
26	.0179	.75	.90625	13	.0897	3.75	3.90625
25	.0209	.875	1.03125	12	.1046	4.375	4.53125
24	.0239	1.	1.15625	11	.1196	5.	5.15625
23	.0269	1.125	1.28125	10	.1345	5.625	5.78125
22	.0299	1.25	1.40625	9	.1495	6.25	6.40625
21	.0329	1.375	1.53125	8	.1644	6.875	7.03125
20	.0359	1.5	1.65625				

TABLE 1-1 BLACK AND GALVANIZED IRON

Galvanized Iron

The most common sheet stock used by the sheet metal worker is galvanized iron, which can be recognized by its bright, spangled appearance, figure 1-3. This stock is made of carbon steel sheets coated with zinc. The steel sheet is usually cleansed in an acid bath and then dipped in a tank of molten zinc.

Figure 1-3 Zinc-spangled surface of a galvanized Iron Sheet
(Courtesy of Inland Steel Co.)

An electroplating process is sometimes used to manufacture a galvanized sheet called *galvanneal* which is a dull, even, gray color. Galvanneal has the ability to hold a painted surface.

A good quality galvanized sheet should last 5 to 10 years under constant contact with water. It can be bent severely without breaking the coating and solders well. It is available in the same sizes as black iron sheets. Specifications are given in table 1-1. Galvanized iron is one of the least expensive of all sheet metals. It is extensively used in ductwork, roof flashings, gutters, tanks, signs, and boxes.

Tin Plate

Tin plate is a silvery, mirrorlike metal made by coating iron or steel plates with pure tin. It is known by trade terms such as IC (pronounced "one c"), IX (pronounced "one cross"), and by the weight of the sheets in a base box. The unit of measure for tin plate is the base box which contains 112 sheets, 14 inches by 20 inches, of approximately 30-gage thickness.

Tin plate is seldom used in the sheet metal shop today. It is limited to occasional odd jobs. Specifications can be found in table 1-2.

Base Wt. Lb.	Trade Term	Approx. U.S. Gage	Approx. Thick. In.	Lb. per Sq. Ft.	Base Wt. Lb.	Trade Term	Approx. U.S. Gage	Approx. Thick. In.	Lb. per Sq. Ft.
55		38	.0061	.253	155	2X	27	.0175	.712
60		37	.0068	.276	168	3XL	26	.0189	.771
65		36	.0073	.298	175	3X	26	.0197	.804
70		35	.0079	.321	180	DX	25	.0203	.827
75		34	.0084	.344	188	4XL	25	.0211	.863
80		33	.0090	.367	195	4X	25	.0220	.895
85		32	.0095	.390	208	5XL	25	.0238	.955
90		31	.0101	.413	210	D2X	25	.0241	.964
95		31	.0107	.436	215	5X	25	.0246	.987
100	1CL	30-1/2	.0113	.459	228	6XL	24	.0261	1.047
107	1C	30	.0120	.491	235	6X	24	.0269	1.079
118		29	.0133	.542	240	D3X	24	.0275	1.102
123		29	.0138	.564	248	7XL	23	.0284	1.139
128	1XL	28	.0144	.588	255	7X	23	.0292	1.171
135	1X	28	.0152	.620	268	8XL	23	.0307	1.231
139	DC	28	.0156	.638	270	D4X	23	.0310	1.240
148	2XL	27	.0167	.680	275	8X	22	.0315	1.263
150		27	.0169	.688					

TABLE 1-2 TIN PLATE

The terms *coke, charcoal,* and *dairyplate* are often used in referring to tin plate. These are commercial terms which refer to thickness of the coating. Coke plate is a cheaper product that carries a tin coating of from 1 to 2.5 pounds per base box. It is used on furnace fittings and pipe to a very limited extent. Charcoal plate coating varies from 3 to 7 pounds per base box. This type of tin plate is used in the food industry for containers and equipment. Plates with a coating of 7 to 14 pounds per base box are called dairyplate. As the name implies, this tin plate is used for milk cans and other dairy equipment.

Terne plate has a coating of lead and tin in a mixture of 75-percent lead and 25-percent tin. The weight of the coating varies from 8 to 40 pounds per double box (112 sheets, 20 by 28 inches). Terne plate is used in roofing, tanks, benches, and cabinets. It is very satisfactory for finishing with acid resistant paint.

Stainless Steel

Stainless steel is a high grade steel to which such elements as manganese, silicon, phosphorus, chromium, nickel, and molybdenum have been added. Stainless steel is, therefore, an alloy: a substance composed of at least one metal and another material. It has a silver-chrome appearance and can be recognized by its grained surface texture. There are over 40 types of stainless steel now available, and no one type is always used in the sheet metal shop. Stainless steel stock is identified by three series of numbers: the 200 series, the 300 series, and the 400 series. Each series identifies the alloying elements. The stock most commonly used in the sheet metal shop is type 302. This stock can be used for ornamental work, containers, sinks, counters, and cabinets.

Stainless steel is also designated by a finish number. These finish numbers range from number 1, which is unpolished, up to number 7, which has a mirrorlike finish. Sheet metal shops usually use the number 3 finish on standard stainless steel jobs.

Although stainless steel is so tough that it does not tear or shear easily, it can be worked in the shop. The cost of stainless steel is about seven times the cost of galvanized iron. However, because of its almost complete resistance to corrosion, it lasts indefinitely. Weights and thicknesses can be found in table 1-3.

APPROXIMATE WEIGHTS AND THICKNESS

Thickness Ordering Range, Inches	Gage No.	Approximate Decimal Parts of an Inch	Avg. Wt. per Sq. Ft. in Lbs. for Chrome Nickel–Cold Rolled Alloys	Average Weight per Square Foot in Pounds for Chrome Iron Alloys
.161 to .176"	8	.171875	7.2187	7.0813
.146 to .160"	9	.15625	6.5625	6.4375
.131 to .145"	10	.140625	5.9062	5.7937
.115 to .130"	11	.125	5.2500	5.15
.099 to .114"	12	.109375	4.5937	4.5063
.084 to .098"	13	.09375	3.9375	3.8625
.073 to .083"	14	.078125	3.2812	3.2187
.066 to .072"	15	.0703125	2.9531	2.8968
.059 to .065"	16	.0625	2.6250	2.575
.053 to .058"	17	.05625	2.3625	2.3175
.047 to .052"	18	.050	2.1000	2.06
.041 to .046"	19	.04375	1.8375	1.8025
.036 to .040"	20	.0375	1.5750	1.545
.033 to .035"	21	.034375	1.4437	1.416
.030 to .032"	22	.03125	1.3125	1.2875
..027 to .029"	23	.028125	1.1813	1.1587
.024 to .026"	24	.025	1.0500	1.03
.0199 to .023"	25	.021875	.9187	.9013
.0178 to .0198"	26	.01875	.7875	.7725
.0161 to .0177"	27	.0171875	.7218	.7081
.0146 to .0160"	28	.015625	.6562	.6438
.0131 to .0145"	29	.0140625	.5906	.5794
.0115 to .0130"	30	.0125	.5250	.515
.0105 to .0114"	31	.0109375	.4594	.4506
.0095 to .0104"	32	.01015625	.4265	.4184

TABLE 1-3 STAINLESS STEEL

Aluminum

Because pure aluminum is very soft, it is very seldom used in sheet form. Almost all aluminum sheets are alloys containing copper, silicon, iron, and manganese. Other alloys of aluminum contain not only these basic elements, but also small quantities of magnesium, chromium and nickel. All aluminum alloys consist of at least 90-percent aluminum and can be identified by their whitish appearance and lightness.

Aluminum sheets are designated by an alloy number. These numbers range from 1100 (pure aluminum) to 7000 (very hard). The sheet metal shop uses number 3003 as the common alloy for ductwork, doors, panels, and awnings.

The workability of aluminum sheets depends on the degree of softness of the alloy. Number 3003 aluminum can be compared in ease of forming with galvanized iron.

Copper

Solid sheets of copper are easily recognized by their reddish color. They are gaged by ounces per square foot instead of thickness. For example, 16-ounce copper is of a thickness that weighs 16 ounces for each square foot.

Cold-rolled copper, that is, sheets rolled to the desired thickness while unheated, is the most commonly used copper stock in the sheet metal shop. It can be stretched, shrunk, and hardened by hammering. It also can be soldered easily.

The great advantage of copper is its high resistance to corrosion and its great workability. Due to its high cost, however, it is mostly used in specialty work, such as ornamental roofs, gutters, downspouts, flashings, and hoods. It has a natural beauty that blends with both the exterior and interior of many styles of architecture.

STRUCTURAL MATERIALS

Although sheet stock is by far the most commonly used material, other forms of metal also play a part in the sheet metal trade. Angle iron, band iron, and wire are widely used as reinforcing and strengthening members in both fabrication and installation.

Angle Iron

Angle iron is a low carbon steel which varies in thickness from 1/8 to 1/2 inch, in width from 1/2 to 2-1/2 inches, and in length from 12 to 40 feet. Angle iron is used for braces, hangers, reinforcing edges and pipe, and for making connections.

Band Iron

Band iron is a low carbon steel which varies in thickness from 1/16 to 1/4 inch, in width from 3/8 to 4 inches, and in length from 12 to 20 feet. Band iron is used for braces and hangers and reinforcing edges.

Wire

Wire is obtainable in 100-pound rolls in a variety of sizes and metals. The more expensive wires - aluminum, copper, and stainless steel - are used occasionally on specific jobs, but common types of wire are steel and iron. Steel and iron wire are usually coated with zinc, tin, or copper to prevent rust and facilitate soldering. Coated wire is commonly referred to as galvanized, tinned, or coppered wire.

The recommended gage for determining wire size is the U.S. Wire Gage and should be used when ordering.

Wire is used for handles and edges on many sheet metal projects, figure 1-4.

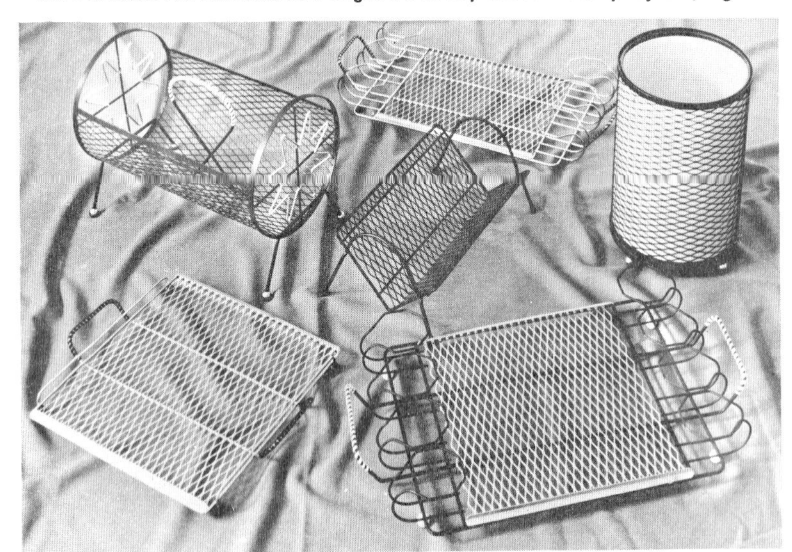

Figure 1-4 Some uses of wire in consumer products

HOW TO SELECT STOCK

1. Refer to the job sheet or to the bill of material on the drawing. Note the kind, size, and shape of material specified.
2. Select the kind of material from the stock rack in accordance with the specifications by noting the external appearance or markings on the material.
3. Select the size and shape of the material in accordance with the specifications on the job sheet or drawing. Band and angle iron can be measured with a scale or rule. The thickness of sheets can be measured with a U.S. Standard Gage or a micrometer and compared with table 1-1, Black and Galvanized Iron, and table 1-2, Tin Plate.

Make sure there are no burrs on the edge of the sheet before measuring the thickness.

4. Select the smallest piece of material necessary to do the job. (See Unit 25, Patterns.)

If several pieces are needed, select a sheet so that the pieces can be cut in the most economical manner.

5. After cutting the stock needed, return the remaining stock to the proper rack.

Place the material of the same size in the proper space in the rack. Place the sheets so that they will not buckle.

SUMMARY REVIEW

A. Place the answers to the following questions in the column to the right.

1. List six types of sheet materials used in the sheet metal trade. 1. ______________________
____
____
____
____
____

2. What are the two general classifications of materials? 2. ______________________
____

3. List three types of metal sheet surfaces available. 3. ______________________
____
____

4. List the standard sheet sizes of black iron and galvanized sheets. 4. ______________________
____
____

5. What is the weight of a bundle of black iron sheets? 5. ______________________

6. What is the life of galvanized sheets exposed to constant contact with water? 6. ______________________

7. What is the greatest advantage of galvanized sheet metal? 7. ____________________
8. List four uses of galvanized iron sheets. 8. ____________________

____

____

____

9. How many sheets are in a base box of tin? 9. ____________________
10. What is the size of the sheet in a base box of tin? 10. ____________________

B. Insert the correct word in the following sentences.

1. Tin Plate is______________ used in the sheet metal shop today.
2. The ability of stainless steel to______________ is its greatest asset.
3. Stainless steel is distributed according to a ______________ number.
4. Pure aluminum sheets are seldom used due to their______________ .
5. The series number______________ stainless and______________ aluminum are commonly used in the sheet metal shop.

C. Underline the correct word in each of the following sentences.

1. Aluminum alloy sheets can be easily recognized by their (dull, gray, red, whitish) appearance.
2. Copper is mainly used today for (common, occasional, specialty) work.
3. Copper can be stretched, shrunk, or hardened by (heating, cooling, hammering).
4. Heavy duct hangers should be made of (sheet metal, wire, angle iron).
5. The raw edge on a container can be eliminated by using (wire, bars, tees).

UNIT 2 METAL PROPERTIES

OBJECTIVES

After studying this unit, the student should be able to

- List the common properties of metals.
- Define metal terms.
- List causes of corrosion and methods of preventing corrosion.

METAL PROPERTIES

As the sheet metal worker uses various types of metals, he should have a general knowledge of their properties. The manufacturer expresses these properties in the terminology of the mill.

Ductility: the ability of metal to permit a change in shape without fracturing. A ductile metal stretches or bends without breaking. Examples are soft iron, copper, and mild steel.

Elasticity: the ability of a metal to hold a load without producing permanent deformation. High carbon steel which is used in springs and bumpers is elastic.

Hardness: the ability of a metal to resist penetration, wear, or cutting action. Steel ball bearings and roller bearings are made of hard metal.

Malleability: the ability of a metal to form when subjected to rolling or hammering. Copper is a malleable metal because it can be hammered, bent, rolled, and twisted without cracking or breaking.

Shortness: the inability of a metal to withstand shattering – commonly called brittleness. Cast iron, very hard steel, and glass are such materials. Files are brittle tools which the sheet metal worker uses.

Strength: the ability of a metal to hold a load under compressive or tensile stress without deformation. Steel is strong, and lead is weak.

MANUFACTURING PROCESSES

All metal, whether in sheets, bars, rods, or angle iron, must go through a manufacturing process that should be familiar to the sheet metal worker. Some of the more common processes are defined as follows.

Alloying: the forming of a material with metallic properties composed of two or more elements of which at least one is a metal. Steel is an alloy because it is composed of iron and carbon.

Annealing: the heating and controlled cooling of solid material for the purpose of removing stresses. Annealing changes the properties of the material by making it softer and more ductile.

Burning: the heating of a metal to a temperature that causes permanent damage to the metal. If a drill is sharpened on a grinding wheel until the tip turns a bluish color, the metal becomes soft and loses its cutting edge.

Cold Finishing: the changing of the shape, or reducing the cross section of a metal while cold. This process is also called cold rolling, drawing, and cold working. Practically all the work a sheet metal worker does on metal is cold work. Cold working causes the metal to become hardened and brittle. If the metal is cold worked too much--if it is bent too sharply, hammered too much or bent back and forth at the same place too often--it cracks or breaks.

Drawing: the pulling of metal though a die. Wire is made by this process.

Forging: the shaping of a piece of metal by hammering or pressing while it is hot. The point of a soldering iron is forged by the sheet metal worker.

Hot Working: the mechanical working of a metal while it is heated to a temperature that causes a structural change. The process is also called hot rolling. Black iron can be cold or hot rolled.

Hardening: the heating and quenching of metal, particularly steel, to increase hardness. Other metals, such as copper, are work hardened by hammering, rolling, and pressing.

Figure 2-1 Forty-five tons of cold-rolled steel in a coil nearly three and a half miles long.
(Courtesy of Inland Steel Co.)

CORROSION

Corrosion, the gradual eating away of metal by chemical action, is caused by exposing metal to fumes, water, acids, and moist air (especially moist salt air). The corrosion of iron and steel is called *rusting,* which is the formation of iron oxide (iron and oxygen) on the surface of the metal. Iron oxide has a reddish-brown appearance. Rusting eventually weakens and destroys the metal. If caught in time, however, rusted metal can be repaired with little or no loss in strength. The rust may form evenly over the surface or occur only in spots.

Corrosion Prevention

It is very important to prevent corrosion since some authorities estimate that 25 percent of the iron and steel produced each year is destroyed by corrosion. Corrosion can be reduced by using grades of iron and steel which contain few impurities, by adding special materials such as small amounts of copper to the metal itself, and by using coatings such as paint, enamel, tin, lead zinc, and cadmium to protect the surface. If the metal is worked so much that surface cracks appear or the protective coating flakes, corrosion takes place.

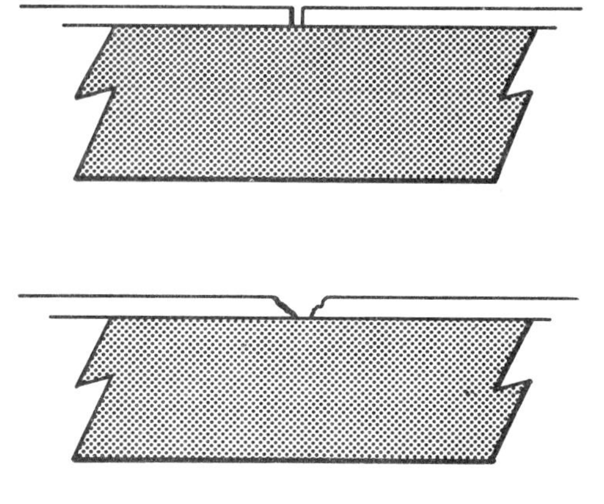

Figure 2-2 Corrosive action at a scratch on a galvanized surface. The sacrificing of the galvanize by galvanic action serves to protect the base metal.

Zinc is used to coat iron and steel because it does not corrode. When a scratch penetrates through the coating to the base metal, zinc sacrifices itself to protect the surface, figure 2-2. Iron exposed to fumes or salt air, however, corrodes rapidly even when galvanized and must be painted. Paint does not adhere well to the surface of new galvanized iron. It is preferable to weather season new galvanized iron one year before painting, or until it has a dull appearance. If weather seasoning is not possible, the metal should first be *killed,* that is, wet with a diluted (5 percent) phosphoric acid solution, allowed to dry, and then rinsed with clear water. Such paints as Metal Treat® and Metal Etching 1616® and phosphoric acid base can be used as a primer. *Bonderizers,* patented phosphate solutions used to coat steel for protection against corrosion, can be applied after priming. Some of the more common bonderizers are X-1-M 400 Flash Bond®, Zin-x-ide 360®, and MZP®. At the present time, there is no guaranteed, nonpeeling paint for galvanized iron on the market.

Metalizing is the spraying of molten zinc on carefully prepared steel surfaces to protect a fabricated structure from corrosion. It is especially useful on large structures, such as bridges, industrial buildings, trestles, tanks, steel piling, ship hulls, and superstructures.

SUMMARY REVIEW

A. Place the answers to the following questions in the column to the right.

1. List the common properties of metals. 1. ______________________
____
____
____
____
____

2. List three metals that are considered to be ductile. 2. ______________________
____
____

3. List two uses of steel which require a high degree of hardness. 3. ______________________
____

4. What would happen to a steel file if it received a sharp blow? 4. ______________________

5. What material would be most suitable to make a hand-hammered ornamental vase? 5. ______________________

B. Insert the correct word in each of the following sentences.

1. The ability of a metal to hold a load without deformation is called its ______________.
2. Steel is an example of (a,an) ______________ because it is composed of iron and carbon.
3. ______________ makes a metal softer and more ductile.
4. Practically all the work a sheet metal worker does on metal is ______________.
5. If the point of a scratch awl turns bluish in color while being sharpened on a grinder, it is said to be ______________.

C. Underline the correct word in each of the following sentences.

1. Steel can be hardened by heating and (hammering, rolling, quenching).
2. Corrosion is caused by exposing a metal to (oil, grease, gasoline, water).
3. The common name for corrosion of iron and steel is (rot, iron oxide, rust).
4. Corrosion can be reduced by covering the metal surface with (paper, paint, cloth).
5. It is best to paint a galvanized metal surface when it appears (bright, glossy, dull, spangled).

UNIT 3 *GAGES*

OBJECTIVES

After studying this unit, the student should be able to

- Define the term gage.
- Identify the methods of determining sheet metal thickness.
- Use the U.S. Standard Gage.

In the sheet metal trade, the word *gage* refers to devices for measuring the thickness of sheets and devices which act as stops when marking, cutting, or forming metals. Sheet thicknesses are also designated by a series of numbers called gages. Several systems are used for different kinds of metals. For the measurement of brass and aluminum sheets the Brown and Sharpe Gage is used; for copper, the Stubbs Gage; and for iron and steel sheets, the U. S. Standard Gage.

U.S. STANDARD GAGE

On March 3, 1893 the Congress of the United States passed an act designating the U.S. Standard Gage as the official gage for iron and steel sheets. The numbers on this gage correspond to the thickness of the sheets. These gage numbers vary from 0000000 for .5 inches thick to 38 for .006 inches thick: the higher the number, the lighter the metal.

Figure 3-1 shows a gage used to measure the thickness of iron and steel sheets. It is a disc-shaped piece of metal having slots of a width to correspond to the U.S. gage numbers. U.S. gage numbers from 0 to 36 are marked on the front of the gage opposite each slot, and the thickness of each gage in decimal parts of an inch is marked on the back.

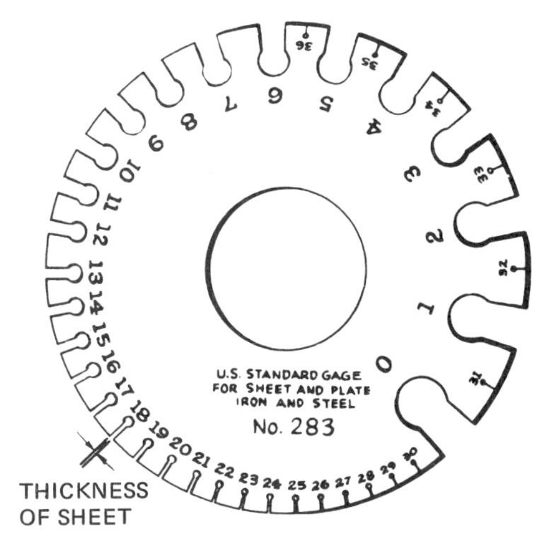

Figure 3-1 United States Standard Gage.

HOW TO USE THE U.S. STANDARD GAGE

1. Select a piece of sheet metal of unknown thickness.
2. Adjust the U.S. Standard Gage until the metal edge slides exactly into the slot.

CAUTION

Do not force the metal tightly into the slot, and do not allow the metal to fit loosely in the slot.

3. Read the gage number (thickness) corresponding to the properly fitted slot.

SUMMARY REVIEW

A. Place the answers to the following questions in the column to the right.

1. What gage system is used to measure the thickness of copper?   1. ________________

2. List three different gage systems used for measuring metal thickness.   2. ________________

____

____

3. What is the thickness range for the U.S. Standard Gage?   3. ________________

4. What is the gage number range marked on the face of the U.S. Standard Gage?   4. ________________

5. Obtain a U.S. Standard Gage and three different thicknesses of sheet metal. Measure and record the gage number and thickness of each.

5.	Number	Thickness
a.	______	______
b.	______	______
c.	______	______

B. Insert the correct word in each of the following sentences.

1. The designation of the thickness of a sheet of metal by a number is called the ____________ .
2. Aluminum sheets are designated according to the ________ ________ system.

C. Underline the correct word in each of the following sentences.

1. A gage can be classified as a (disc, plate, stop).
2. A U.S. Standard Gage can be used to measure the gage of (angle iron, band iron, plate, wire).

UNIT 4 RULES AND TAPES

OBJECTIVES

After studying this unit, the student should be able to

- Identify and describe the types of rules and tapes used by the sheet metal worker.
- Select and use a rule properly.
- Use the correct tape to make inside, outside, and circumference measurements.

The common measuring tools used by the sheet metal worker are the zig-zag or folding rule, the circumference rule, the push-pull steel tape, and the 50- or 100-foot steel tape.

All linear measurement tools are calibrated in graduations of 1/16, 1/8, 1/4, 1/2 inch, and full inches. The angular measurement tools are calibrated in degrees or one-half degrees (30 minutes).

Zig-Zag Rule

The zig-zag rule, figure 4-1, is made of wood or steel and aluminum in 3 to 8 foot lengths, and it may have outside or inside readings. The sheet metal worker prefers to use the inside reading rule because it lies flat and close to the work. This increases accuracy in layout work. The carpenter uses the outside reading rule. For convenience in reading, many rules use red numbers for the foot marks and black for the inch divisions. Likewise, standard stud spacing (16 inches) is shown in large numbers.

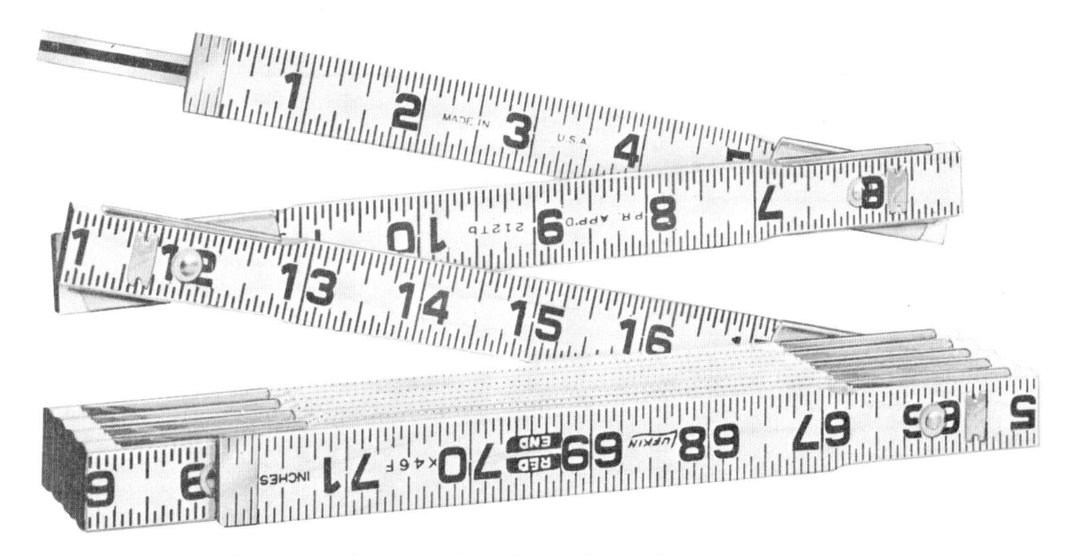

Figure 4-1 Inside reading, 6-foot, heavy-duty zig-zag rule.
(Courtesy of Cooper Industries.)

HOW TO USE THE ZIG-ZAG RULE

1. Hold the rule in one hand and unfold the sections one at a time until the desired length is obtained.

Place a drop of oil at each joint to avoid breaking the rule when unfolding.

2. Place the unfolded rule flat on the sheet metal surface.
3. Adjust the end to the starting point of measurement.
4. Mark the desired length with a scratch awl.

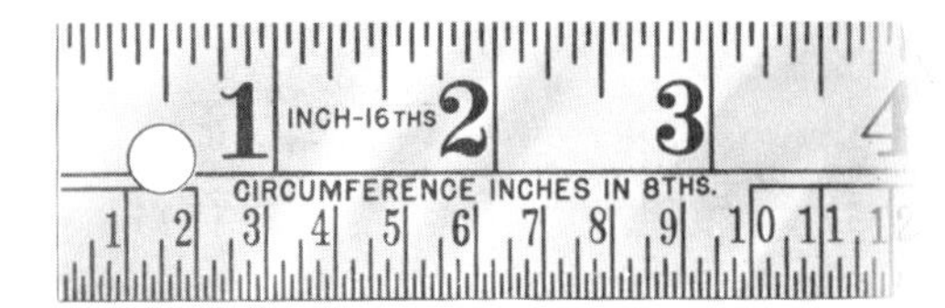

Figure 4-2 Tinner's steel circumference rule.
(Courtesy of Niagara Hand Tools.)

Circumference Rule

The circumference rule, figure 4-2, is a standard rule used by all types of sheet metal workers. It is used not only in measurement, but also in the layout of straight lines on flat and curved surfaces. This rule comes in lengths of 36 and 48 inches. Its width is 1 1/4 inches. On one edge the inch scale is graduated in sixteenths, and on the opposite edge the circumference lengths are graduated in eighths. The reverse side of the rule has formulas for calculating circumferences and areas and tables for laying out measures and cans of various capacities.

HOW TO USE THE CIRCUMFERENCE RULE

1. Use the rule for straight line measurement as outlined for the zig-zag rule.
2. For line layout, place the edge of the rule at two measured points and draw a line with scratch awl.
3. To measure a circumference, locate the diameter on the inch scale and read directly below it on the other edge.
4. Lay off the circumference length, starting from the zero end of the rule to the length located in step 3.

The Push-Pull Steel Tape

The push-pull steel tape is manufactured for the convenience of the worker. It fits easily into a pocket or tool kit. Push-pull tapes, figure 4-3, can be obtained in lengths of 6, 8, 10, and 12 feet and widths of 1/4, 1/2, 5/8 and 3/4 inches. Some tapes have spring rewinds, but others must be hand operated. The end of the tape is usually fitted with a steel hook for ease in taking measurements. The inch scale, usually on both edges of one side only, is customarily in sixteenths. Some tapes, however, are graduated in thirty-seconds for the first six inches. The winding case has a flat bottom of a definite length for accurate inside measurements. Some have a small red pointer to indicate the direct reading. The flexibility of this tape makes it suitable for measuring curved and round surfaces.

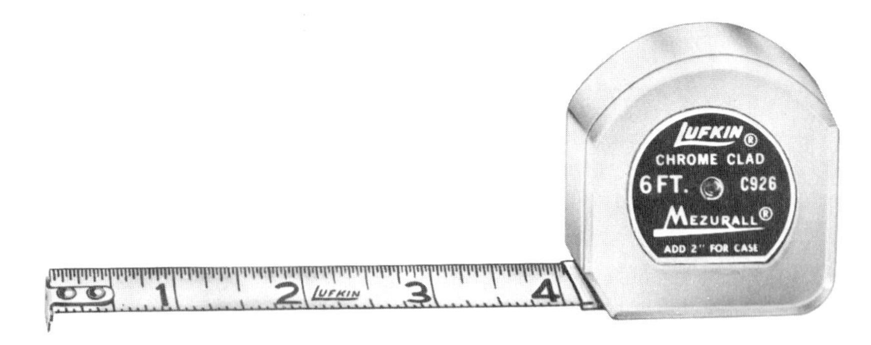

Figure 4-3 Push-pull steel tape.
(Courtesy of Cooper Industries.)

HOW TO TAKE AN INSIDE MEASUREMENT

1. Put the end of the tape against one side of the opening. Uncoil the tape across the opening toward the opposite side.
2. Hold the tape at the starting position and continue extending it until the outer edge of the case butts against the opposite side of the opening.
3. Add the width of the tape case - usually 2 inches - to the reading. Another type of pocket rule allows direct reading of the inside measurement by a small red pointer as shown in figure 4-4.

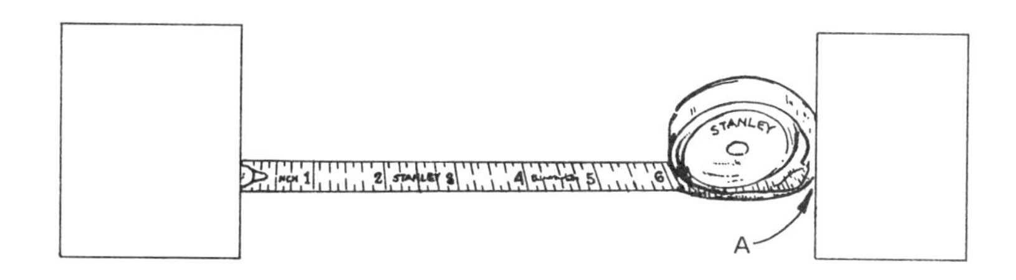

Figure 4-4 Push-pull rule with pointer (a) for direct reading of inside measurement.

HOW TO TAKE AN OUTSIDE MEASUREMENT

1. Pull the tape from the case until it projects far enough to permit measuring the required distance.
2. Hook the end of the tape over the end of the object to be measured, figure 4-5.
3. Read the measurement on the tape according to the graduation which lines up with the point being measured.

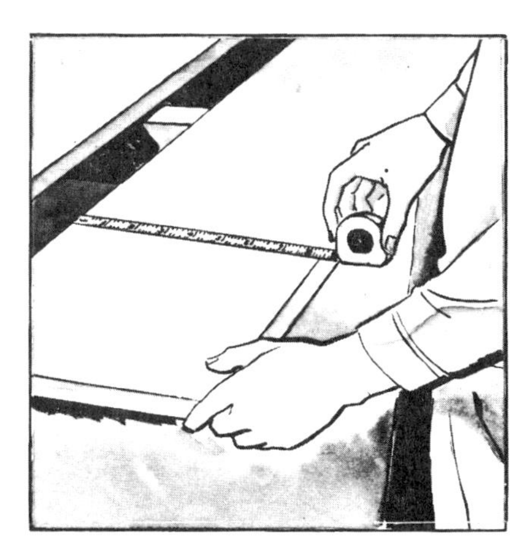

Figure 4-5 Taking an outside measurement.

HOW TO MEASURE THE CIRCUMFERENCE OF A CYLINDER

1. Pull the tape out a short distance and hold the end close to the cylinder as shown in figure 4-6.
2. Pull the tape around the cylinder as if wrapping it. Hold the end in a fixed position as you uncoil the tape from the case. Completely encircle the cylinder with the tape so that it goes beyond the point where it first made contact with the surface of the cylinder.
3. Note which graduations are in line where the tape crosses at the starting point. If the 1-inch and 7-inch graduations are in line, figure 4-6, the circumference is found by subtracting the 1-inch reading from the 7-inch reading. Therefore, 6 inches is the circumference measurement.

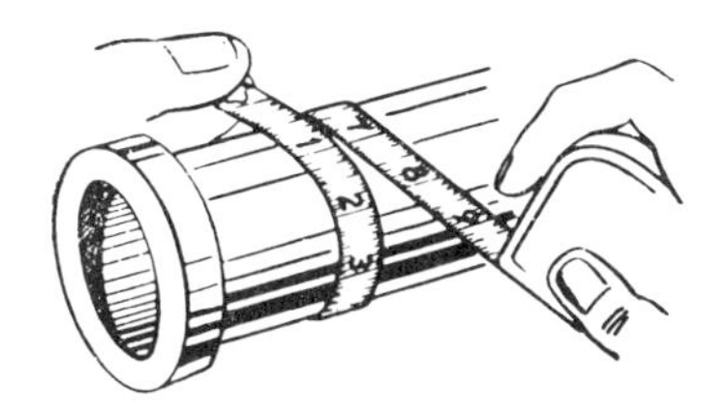

Figure 4-6 Measuring a circumference with a push-pull rule.

Steel Tapes

Steel tapes are used to measure lengths up to 50 and 100 feet. They are made in many types to suit particular needs. The tape is made of flexible spring steel, stamped in graduations of feet, inches, half-inches, quarters, eighths, and, in some instances, sixteenths of an inch.

Cloth or wire-woven tapes may be used when only a reasonable degree of accuracy is needed. They should not be used when measurements must be precise because of their tendency to shrink or stretch. When precise measurements of long lengths are required, a steel tape is preferred.

Two types of tapes are illustrated in figures 4-7 and 4-8. Note that each tape has a ring on the end so that the tape can be anchored over a nail.

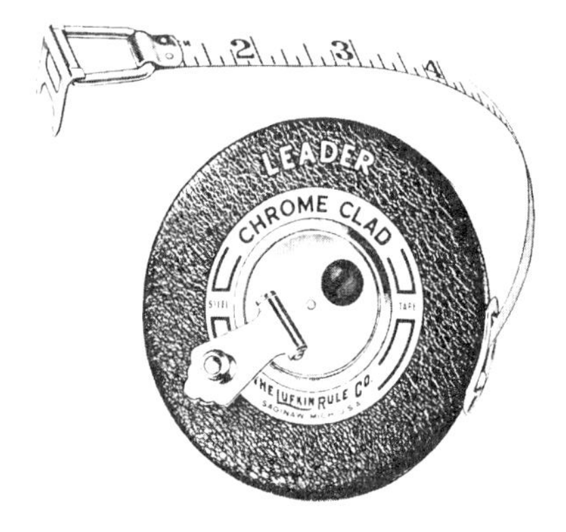

Figure 4-7 Steel tape.

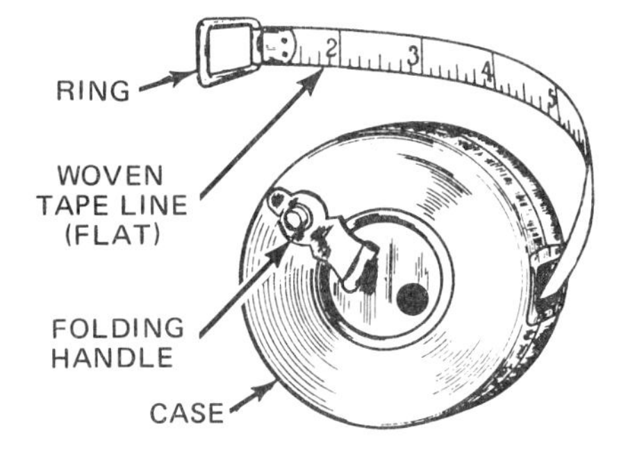

Figure 4-8 Woven tape.

Some steel tapes contain a hook on the ring for anchoring the tape at the ends of boards. Several types of hooks used on tapes are illustrated in figure 4-9.

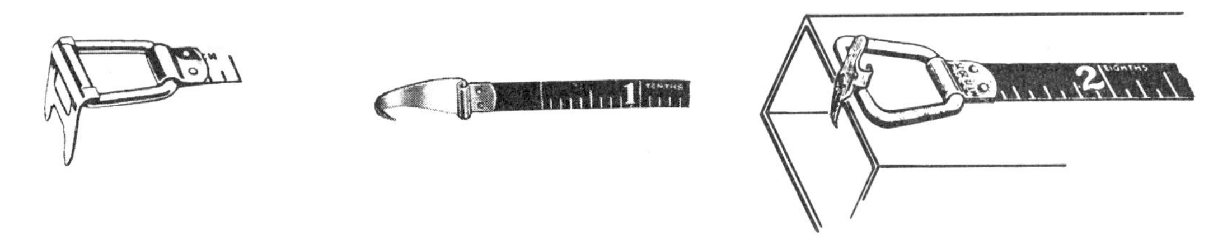

Figure 4-9 Types of hooks used on measuring tapes.

The steel ribbon is uncoiled from the case by pulling outward on the ring in the direction in which the measurement is to be taken, see *A*, figure 4-10. Pulling as in *B* damages the tape. The winding handle is usually opened by pressing on the center of the opposite side of the case. The ribbon is recoiled into the case by turning the handle clockwise.

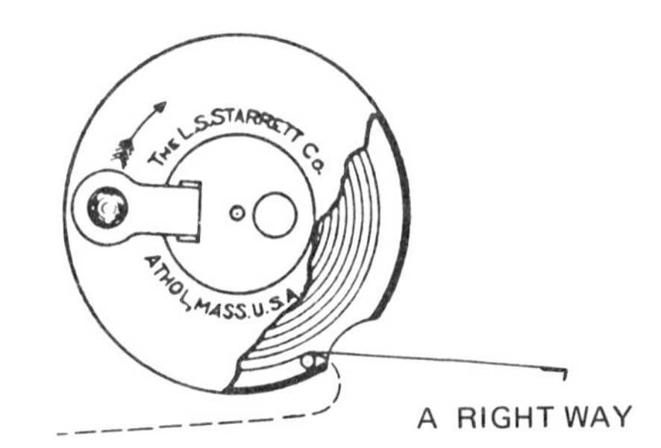

Figure 4-10 Right and wrong methods of withdrawing tape from case.

At times, tapes may pull hard or stick. Tapping the side of the case against a flat surface usually frees the tape. Do not step on the tape or let it become twisted when it is uncoiled. Short kinks crack the tape. To prevent rust, uncoil the tape and wipe it with an oily cloth. This treatment should be given frequently, especially after using the tape in damp weather.

HOW TO MEASURE DISTANCE WITH A 50- OR 100-FOOT STEEL TAPE

1. Secure the ring end of the tape to the point from which the measurement is to be taken. This may be done by driving a nail at the point and then adjusting it so that when the ring of the tape is slipped over the nail, the outside edge of the ring is directly over the point from which the measurement is to be taken, figure 4-11A. If the hook is used, it may be applied as shown in figure 4-11 B or C.

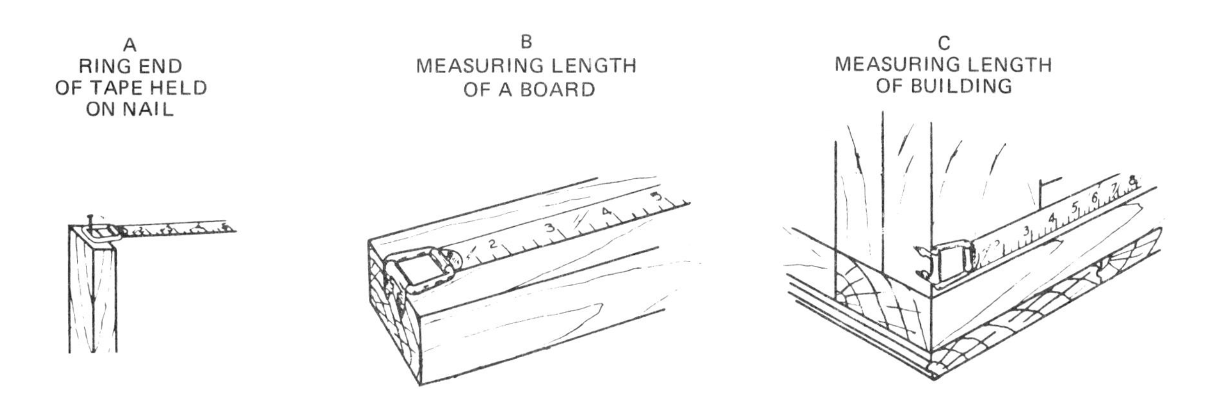

Figure 4-11 Applications of steel tape.

2. Uncoil the tape from the case in the direction of the point to which measurement is to be made. Be sure that the tape lies flat and is free from kinks.
3. Pull the tape taut to the desired point of measurement. Mark the desired point directly opposite the graduation on the tape.
4. For measuring the distance between two points, read the graduations on the tape by noting which line on the tape coincides with the point of measurement.
5. Recoil the tape into the case after the measurements have been taken by opening the winding handle and turning it clockwise.

SUMMARY REVIEW

A. Place the answers to the following questions in the column to the right.

1. What type of zig-zag rule is most useful to the sheet metal worker? — 1. ____________
2. How is the inch scale graduated on the — 2. a. ____________
   a. zig-zag rule — b. ____________
   b. circumference rule
3. Which rule should be used to lay out a straight line 21 inches long--a zig-zag rule or a circumference rule? — 3. ____________

4. Using a circumference rule, list the circumference for the following diameters.
   a. 1 1/2 inches
   b. 2 1/4 inches
   c. 3 inches
   d. 33 3/4 inches
   e. 35 1/8 inches

4. a. ____________
   b. ____________
   c. ____________
   d. ____________
   e. ____________

5. On which rule can the measurements for a 1-quart container be found?

5. ____________

6. List three types of surfaces that can be measured with a steel tape.

6. ____________
   ____
   ____

7. For which kind of measurement is the hook on the end of the tape useful?

7. ____________

8. How is the inch scale usually graduated on
   a. the push-pull tape
   b. the 50- and 100- foot steel tape

8. a. ____________
   b. ____________

9. Why are cloth and wire-woven 50- and 100- foot tapes not recommended for precise measurements?

9. ____________

B. Insert the correct word in each of the following sentences.

1. The right way to uncoil a 50- or 100-foot steel tape is to pull ____________ on the ring.
2. To prevent the steel tape from rusting, wipe it with an __________ cloth.
3. Short kinks in a steel tape cause it to _______.
4. The purpose of the hook or ring on the end of a steel tape is to ___________ that end.

C. Underline the correct word in each of the following sentences.

1. A (folding rule, circumference rule, push-pull steel tape) is used to measure the inside opening of a ventilation duct accurately.
2. The worker would select a (zig-zag rule, push-pull steel tape, 100-foot cloth tape, 100-foot steel tape) to measure the length of a duct run in a large shop.

UNIT 5 LINE MARKERS AND INDENTATION MARKERS

OBJECTIVES

After studying this unit, the student should be able to

- Identify and describe the various line markers and indentation markers used by the sheet metal worker.
- Explain the differences between the prick punch and center punch.
- Use the markers correctly.

Marking tools are a basic necessity to the sheet metal worker. These tools must be able to scribe a line, circle, or arc permanently on a metal sheet so that the worker can follow it accurately with his cut. Prick punches and center punches are markers which permanently locate points, centers, and pattern outlines. Markers are made with hardened steel points for wearing quality and accuracy. Other marking devices, such as the soapstone and felt-tipped pen, are used in installation work when marking is needed only for identification purposes.

The Scratch Awl

Scratch awls, figure 5-1, are available in lengths ranging from 5 1/2 to 9 inches. Handles are made of wood, steel, or plastic. The pointed blades are long and cylindrical, and measure from 1/8 to 1/4 inches in diameter. They are heat treated to hold a needle point. The points can be sharpened on a grinding wheel when necessary.

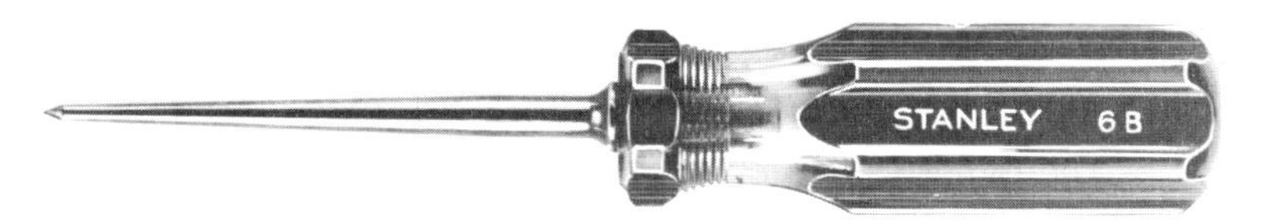

Figure 5-1 Scratch awl. (Courtesy of Stanley Tools.)

HOW TO USE A SCRATCH AWL

1. Hold the scratch awl as if holding a pencil.
2. Mark the line location at two points on the metal surface.
3. Place a straightedge on the location points.
4. Draw a line with the scratch awl along the straightedge.

Dividers

Dividers, figure 5-2, are available in 6-, 8-, and 10-inch lengths. They are adjusted to the desired dimension by a wingnut lock or screw. The points are heat-treated steel to hold sharpness. Resharpening can be done on the grinding wheel. Dividers are used for scribing circles or arcs as large as 48 inches in diameter. Other uses are line, arc, and circle division, and line transfer on a layout.

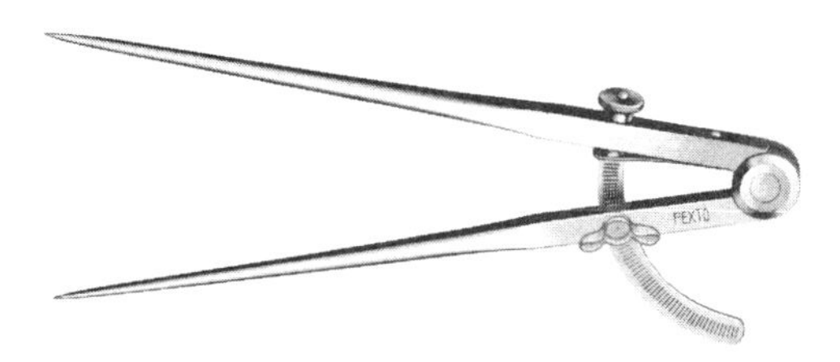

Figure 5-2 Wing dividers. (Courtesy of Pexto.)

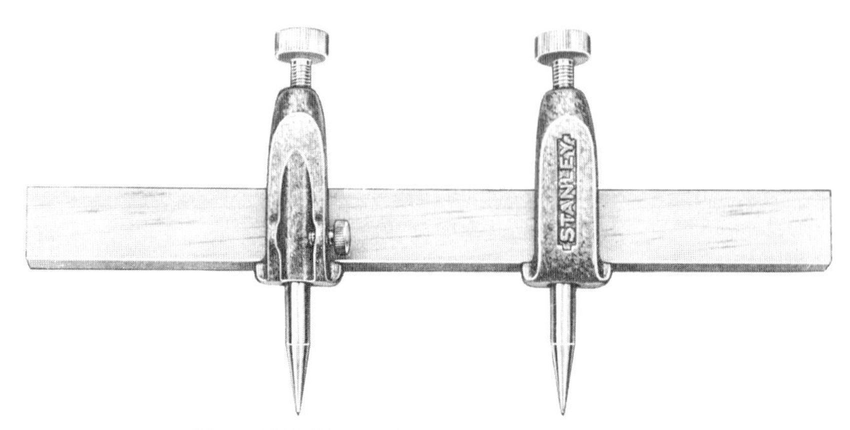

Figure 5-3 Trammel. (Courtesy of Stanley Tools.)

The Trammel

When circles or arcs larger than 48 inches must be scribed on a surface, a trammel is used, figure 5-3. The trammel can be obtained with an adjustable sliding mechanism, and individual trammel points can be fastened to a rod of any desired length.

HOW TO USE A DIVIDER OR TRAMMEL

1. Locate the center of the circle or arc on the stock.
2. Indent the center with a prick punch or scratch awl.
3. Adjust the points of the dividers or trammel to the desired radius and secure.
4. Place one point in the center location as a pivot.
5. Rotate the other point by hand to scribe a circle.

The Marking Gage

The most common straight line marking gage used by the sheet metal worker is shown in figure 5-4. This gage can be purchased or made from a small piece of sheet stock. It is calibrated for standard seam and cleat allowances which are indicated by the figures either tamped or marked on the face. There are other types of gages with points adjustable to any desired dimension, but they are seldom used in the sheet metal trade.

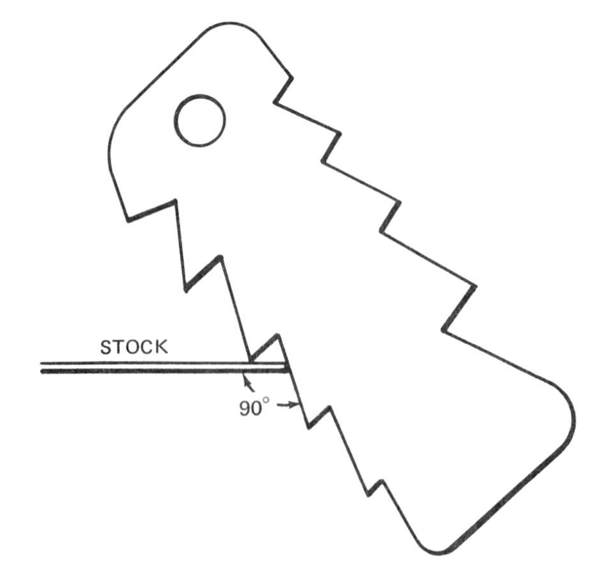

Figure 5-4 Marking gage.

HOW TO USE THE MARKING GAGE

1. Place the edge of the sheet to be marked on the bench so that it overhangs slightly.
2. Select the dimension to be used.
3. Place the shoulder of the selected dimension perpendicular to the edge of the stock.
4. Move along the edge to scribe a line with the point of the gage.

The Prick Punch

The prick punch, figure 5-5, is made of tool steel which is hardened and drawn to hold a long, sharp, point. The conical point is ground to an included angle of 30 degrees like that of the scratch awl. The prick punch is very useful for transferring patterns on sheet metal surfaces and for punching hole centers.

Figure 5-5 Prick punch. (Courtesy of Stanley Tools.)

HOW TO USE A PRICK PUNCH

1. Scribe the points to be marked on the stock.
2. Hold the prick punch vertical to the stock surface with the point on the scribed location.
3. Tap the head of the punch lightly with a hammer.

The Center Punch

Center punches, figure 5-6, are very similar to the prick punch. They are also made of tool steel and are available in the same lengths and diameters as prick punches. The center punch differs only in respect to its point which is blunter than that of the prick punch and has an included angle of 90 degrees. The purposes of this tool are to mark hole centers and enlarge prick punch marks for drilling.

Figure 5-6 Center punch. (Courtesy of Stanley Tools.)

HOW TO USE A CENTER PUNCH

1. Place the center punch so that it is vertical to the hole center location.
2. Tap the head of the punch sharply with a hammer.

CAUTION — If the mark is made slightly off center, correct it by tipping the punch and hitting it towards the correct center.

SUMMARY REVIEW

A. Place the answers to the following questions in the column to the right.

1. Identify three line marker tools used in the sheet metal trade. — 1. ______

______

______

2. What must a line marker be able to accomplish? 2. ____________________
3. List two marking tools used for identification purposes in installation work. 3. ____________________ ____________________
4. Give two reasons why the points of marking tools are heat-treated and hardened. 4. ____________________ ____________________
5. How are the points on a scratch awl and dividers sharpened? 5. ____________________
6. List two uses for the prick punch. 6. ____________________ ____________________
7. What are the purposes of the center punch? 7. ____________________ ____________________
8. How is an off-center punch mark corrected? 8. ____________________
9. What tools are used to show the hole centers in
   a. an aluminum pattern b. 1/4-inch steel plate 9. a.__________ b.__________
10. List the marking tools needed to lay out the cone pattern in the drawing below on 26-gage galvanized iron. 10. ____________________ ____________________ ____________________

B. Insert the correct word in each of the following sentences.

1. To transfer a metal pattern to a sheet of 16-gage black iron, the ______________ is used.
2. The ________________ is used to lay out the edge allowances for a 1/4-inch double seam.
3. Large arcs are scribed on sheet stock with the __________.
4. The included angle on a center punch point is ground to _____________.
5. The prick punch is held in a __________ position over a center location.

C. Underline the correct word in each of the following sentences.

1. A ventilation fitting calls for a heel radius of 85 inches. To make this layout the (scratch awl, dividers, trammel) is used.
2. The cheek pattern of an offset is to have a 1/4-inch flange for seaming. This can be scribed easily on the pattern edge with the (scratch awl, trammel, marking gage).
3. The prick punch has a (short, blunt, long, cylindrical) point.
4. The difference between a prick punch and a center punch is the (length, diameter, point).
5. A pattern can be transferred easily with the (dividers, scratch awl, prick punch).

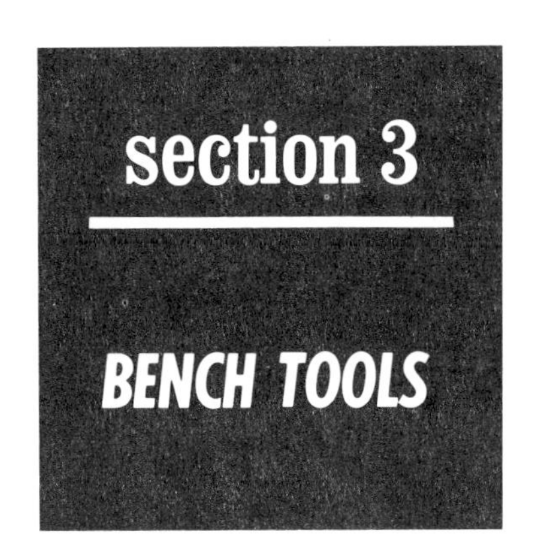

UNIT 6 *SCREWDRIVERS*

OBJECTIVES

After studying this unit, the student should be able to

- Identify and describe the commonly used screwdrivers.
- Select the proper screwdriver for the job.
- Use each screwdriver correctly.

A sheet metal worker is often judged by the way he uses and cares for his tools. In order to gain an appreciation of tools and to be able to select good quality tools, the beginner should become familiar with the construction and materials used in the making of these tools. He must also learn to keep his tools in their proper place and sharpen and repair them when they become worn or broken. With proper care, tools last longer, give better service, and thus are less expensive over a period of time.

USES OF SCREWDRIVERS

Screwdrivers are used primarily for tightening or loosening slotted screws. The sizes and shapes of screwdrivers vary in construction and design, depending upon their use. Some of the screwdrivers used by the sheet metal worker include the standard screwdriver, the square shank screwdriver, the Phillips screwdriver, the ratchet screwdriver, the spiral ratchet screwdriver, and the offset screwdriver.

The Standard Screwdriver

The standard blade screwdriver consists of three parts: the handle, the clamp or ferrule, and the blade. The handle is made to fit the grasp of the hand and is composed of wood, metal, or plastic materials.

The blade is generally made of a good grade of steel, forged, machined to shape, hardened, and tempered. The shank end of the blade is *tanged* (flattened) to prevent turning in the handle. The ferrule fits over the end of the handle and serves to protect the handle and to clamp it tighter to the blade.

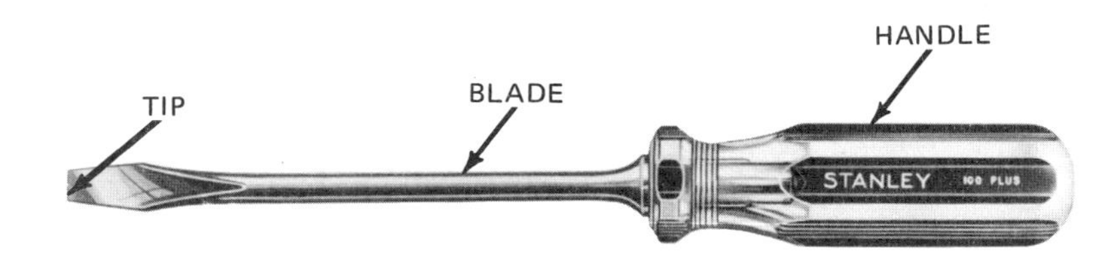

Figure 6-1 Standard blade screwdriver. (Courtesy of Stanley Tools.)

Many sizes and styles of the standard screwdriver are available. The size is determined by the length of its blade. For example, an 8-inch screwdriver has a blade 8 inches long. Screwdrivers with long blades require less effort to drive screws than short ones. They are not suitable, however, for driving small screws because of the danger of twisting the screw to the point that it breaks. Screwdrivers with small tips are intended for use on small screws; and those having wide tips, on large screws. A range of tip widths is necessary for driving screws of different sizes.

The standard screwdriver is used for general purpose work. Its range of sizes permits its application in limited spaces as well as unobstructed areas and for light and heavy driving. The standard screwdriver is not practical when a great many screws are to be driven or when speed is an important consideration.

The Square Shank Screwdriver

The square shank screwdriver, figure 6-2, differs from the standard type only in the shape of the shank portion of the blade. The shank is square so that a wrench may be applied to give added leverage in driving or drawing screws.

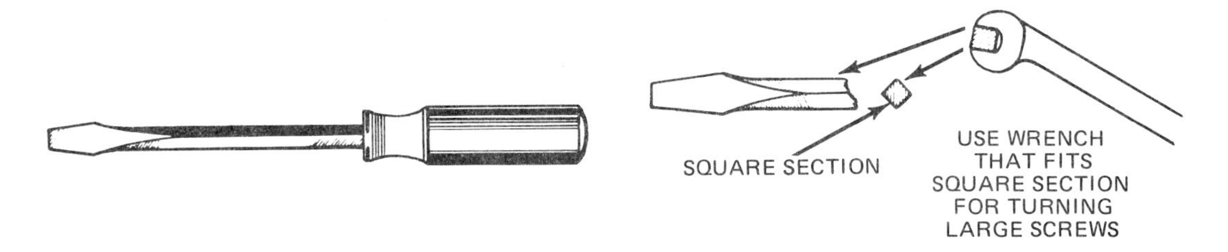

Figure 6-2 Square shank screwdriver.

The Phillips Screwdriver

A Phillips screwdriver is like the standard screwdriver except that the tip of the Phillips screwdriver is shaped like a cross so that it can fit into the slots on Phillips head screws, figure 6-3. The length of the blade signifies the size of the screwdriver.

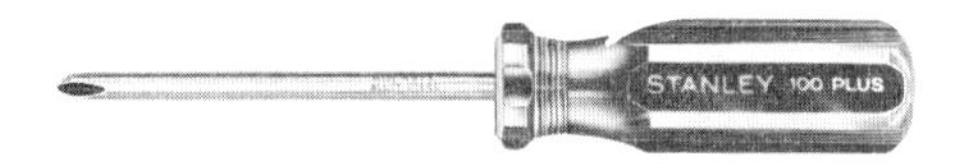

Figure 6-3 Phillips screwdriver. (Courtesy of Stanley Tools.)

Unlike the standard screwdriver, the tip of the Phillips screwdriver is sized according to a particular range of screw sizes. Refer to table 6-1 for the appropriate tips to use for Phillips screws.

Phillips Bit No.	Screw Gage Size
Bit No. 1	Number 4 and smaller
Bit No. 2	Number 5 to 9 inclusive
Bit No. 3	Number 10 to 16 inclusive
Bit No. 4	Number 18 and larger

TABLE 6-1

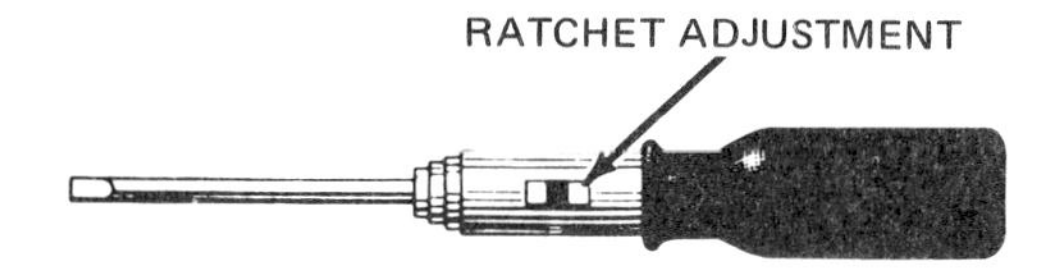

Figure 6-4 Ratchet screwdriver.

Ratchet Screwdrivers

The simple ratchet screwdriver, figure 6-4, has a ratchet device on the ferrule portion of the handle. The ratchet has a wheel and pawl mechanism which allows motion in one direction only. The ratchet permits driving screws at a faster rate and with greater ease. It may be disengaged so that it can be used as a standard screwdriver.

Spiral ratchet screwdrivers, figure 6-5, may be compared with an automatic drill. The turning motion of the screwdriver bit is produced by pushing down on the handle, thereby causing the spiral-groove spindle to revolve. These screwdrivers are available with various lengths of spiral groove spindles. Those with short spindles are better suited for driving short screws.

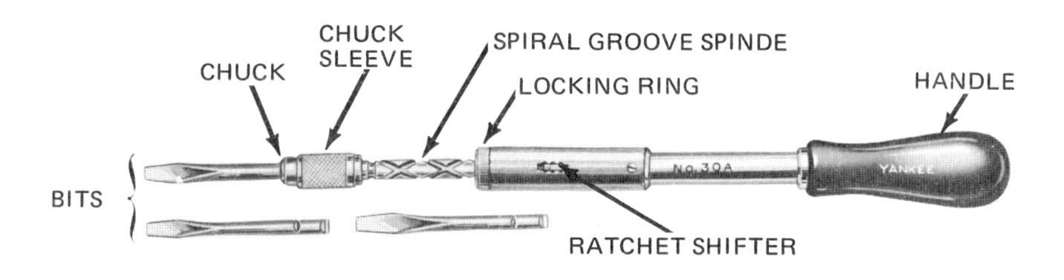

Figure 6-5 Spiral ratchet screwdriver. (Courtesy of Stanley Tools.)

An assortment of bit sizes is usually provided with the screwdriver. The bits are easily interchangeable by adjusting the chuck sleeve.

The spiral ratchet screwdriver is most useful for the rapid driving and drawing of screws and is especially efficient in repetitive production work. It may be used with or without the ratchet in operation.

The Offset Screwdriver

The offset screwdriver, which is made from either octagonal or round stock, can drive and extract screws where there is not enough room to use the standard screwdriver. Both ends of the offset screwdriver are at right angles to the midsection, or shaft, figure 6-6A. The blades on both ends are at right angles to each other. When there is limited turning space, figure 6-6B, the blades may be alternated back and forth until the screw has been driven or extracted.

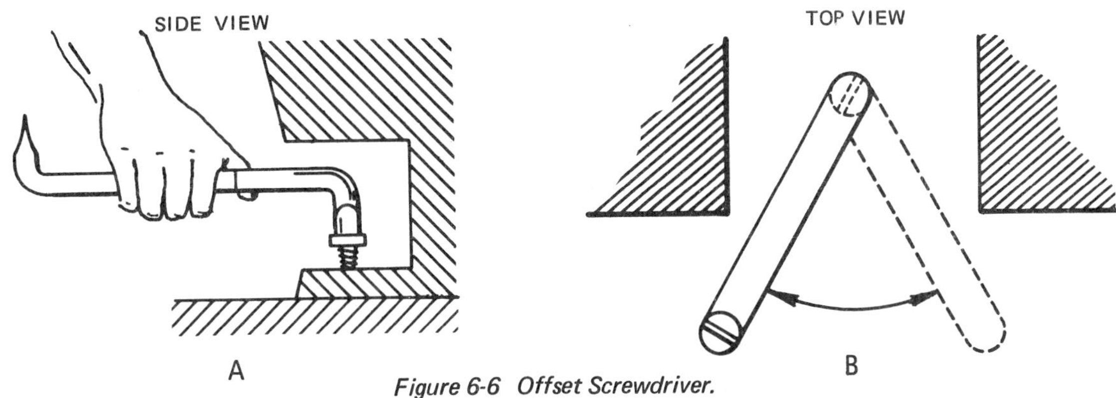

Figure 6-6 Offset Screwdriver.

Figure 6-7 Offset ratchet screwdriver. (Courtesy of Stanley Tools.)

Offset ratchet screwdrivers, figure 6-7, are available for both the standard slotted-head screw and the Phillips head screw. The offset ratchet handle can be fitted with 1/4-inch or 3/8-inch blades for slotted screws, and blades for numbers 0 - 9 and numbers 5 - 16 Phillips screws.

HOW TO SELECT THE PROPER SCREWDRIVER

When selecting a screwdriver, make sure that the tip is in good condition, figure 6-8, and that it fits the screw slot properly, figure 6-9. If the tip is not in good condition, it should be squared and ground to the correct shape. Since the tip has been hardened and tempered, care must be used not to *burn* (overheat) the tip.

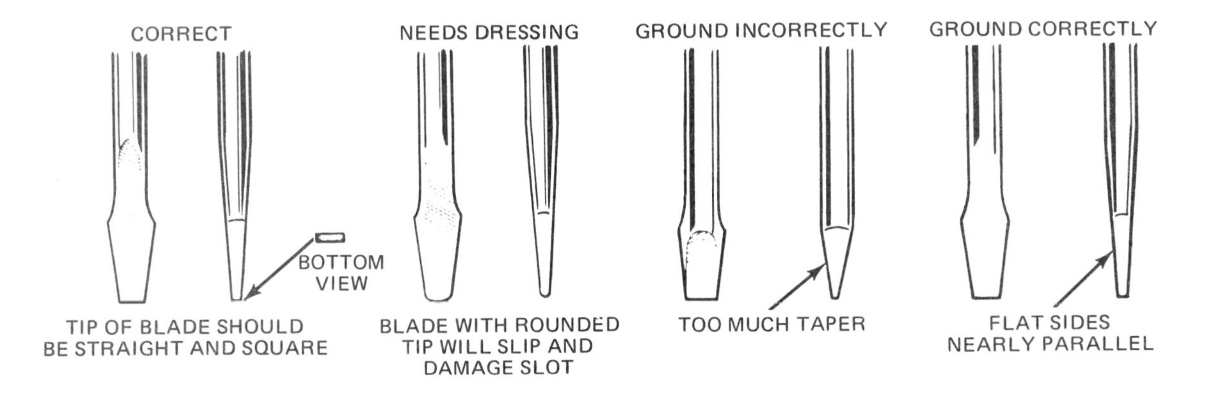

Figure 6-8 Blade tip conditions.

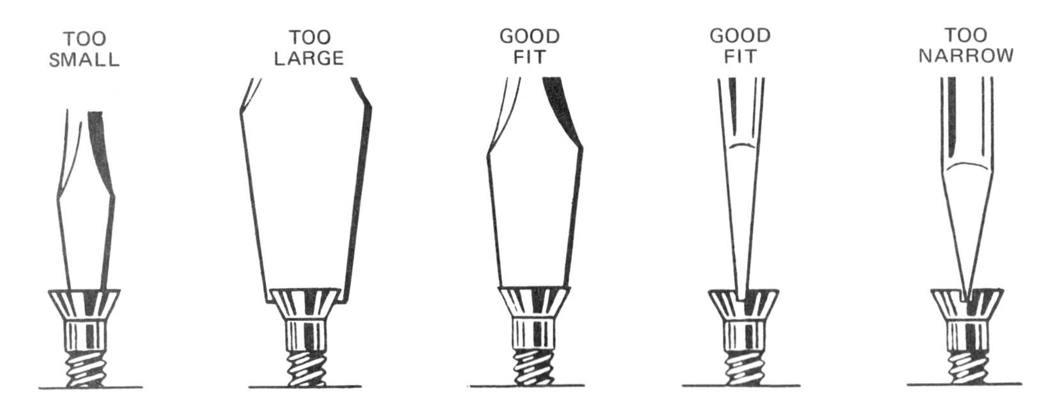

Figure 6-9 Tip selection.

The tip selected should fit the slot snugly and extend the full length, or nearly so, of the slot. If the tip is wider than the slot, the work surface is marred around the screw head as the screw is seated. A blade that is too narrow in width and thickness causes burring of the screw head and often damages the tip itself.

HOW TO DRIVE A SCREW

1. Select a screwdriver of the proper type and size.

2. Place the screw in the clearance hole and while holding it straight with the left hand, center the tip of the screwdriver in the slot and hold it in alignment with direction of the hole.

3. Turn the screw slowly in a clockwise direction, applying enough pressure to drive the screw squarely into the hole.

Keep the fingers near the blade tip away from the underside of the screw head to avoid injury if the screwdriver slips out of the slot.

4. After the screw has been inserted, hold the screwdriver with both hands as shown in figure 6-10. Continue driving the screw with a series of turns by taking a tighter grip on the blade as the handle is turned.

Do not apply more force than is necessary to seat the screw firmly; otherwise the screw head may be twisted off, or the screw threads may be stripped.

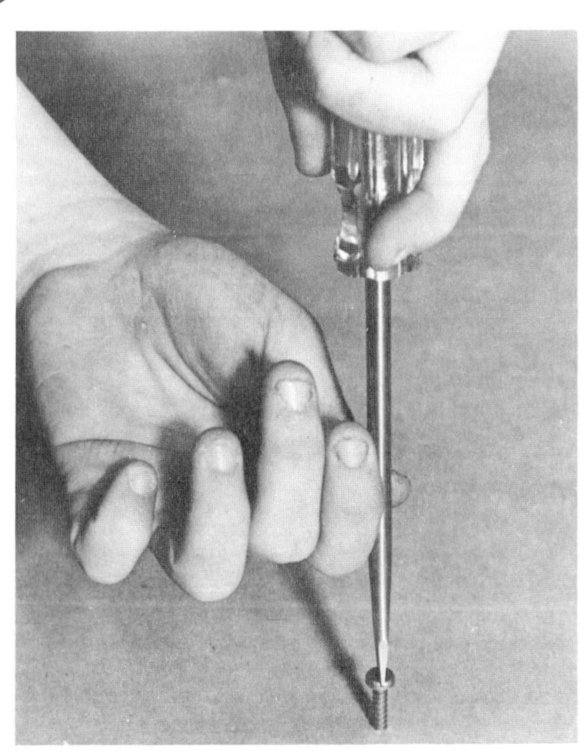

Figure 6-10 Proper position of hands for driving a screw.

5. If the screws are Phillips head, use the same technique described in steps 1 through 4, but exert greater pressure while driving so that the tip does not slip out of the slots.

HOW TO WITHDRAW A SCREW

1. Clean the slot on the screw head so that the tip of the screwdriver seats itself fully.

2. Use the screwdriver in the same way described for driving a screw, but change the direction of rotation to counterclockwise.

3. If the screw is extremely tight and it cannot be withdrawn on the first attempt, try turning it clockwise slightly to loosen it, and then turn it counterclockwise.

4. If a screw binds as it is withdrawn, work it both ways, gradually working it out of the hole.

5. If the screw slot becomes damaged, grasp the head with a pair of pliers after it is part way out and complete the withdrawal with the pliers.

SUMMARY REVIEW

A. Place the answers to the following questions in the column to the right.

1. List three different types of screwdrivers in use. 1. ____________________

   ____

   ____

2. List the three parts of a screwdriver. 2. ____________________

   ____

   ____

3. When is a screwdriver with a long blade used? 3. ____________________
4. How is a screwdriver selected? 4. ____________________
5. When are square shank screwdrivers used? 5. ____________________
6. For what screws is a small tip screwdriver used? 6. ____________________
7. What screwdriver is used to drive a series of screws? 7. ____________________
8. If a screw is recessed in an opening, what screwdriver should be used to drive it? 8. ____________________
9. If the screwdriver blade is too narrow in width and thickness for the screw, what will happen? 9. ____________________
10. Should the worker ever hold an object with one hand and try to drive a screw into it with the other? Why, or why not? 10. ____________________

B. Insert the correct word in each of the following sentences.

1. Screwdrivers with wide tips are intended for use on __________ screws.
2. (A,An) _______ can be used on the square shank of a screwdriver to increase the leverage.
3. The tip on a Phillips screwdriver is shaped like (a, an) __________ .
4. The ____________ screwdriver facilitates driving screws at a faster rate.
5. The ____________ of an offset screwdriver are at right angles to one another.

C. Underline the correct word in each of the following sentences.

1. A screwdriver blade that is too narrow in width and thickness for the slot causes (injury, breakage, burring).
2. If a screw is driven beyond the point of seating firmly, it can be (damaged, marred, twisted off).
3. If a damaged screw slot requires replacing the screw, it should be withdrawn slightly with a screwdriver and then completely with (a wrench, claw hammer, pliers).
4. In driving a series of the same size screws, the (conventional screwdriver, offset screwdriver, ratchet screwdriver) should be used.
5. The tip of the blade of any screwdriver should be (rounded, tapered, straight and square).

UNIT 7 HAMMERS AND MALLETS

OBJECTIVES

After studying this unit, the student should be able to

- Identify and describe the various hammers used in the sheet metal trade.
- Explain the difference between hammers and mallets.
- Use a hammer and mallet properly.

It is essential that sheet metal workers have a variety of hammers. These hammers are made in two types: those with hard heads made of good quality steel forgings and those with soft heads of lead, copper, babbitt, rubber, rawhide, and wood.

Hard head hammers come in many different shapes, depending upon their use. The three common shapes of hammers used by sheet metal workers are the ball peen hammer, the setting hammer, and the riveting hammer.

The Ball Peen Hammer

One end of the head of the ball peen hammer is perfectly ball shaped. This is the peening end. The other end, the face, is slightly curved with chamfered edges. Both the peen and face are polished and tempered to assure safety to the user. This hammer is used for general purposes, and is obtainable in a range of weights from 2 ounces to 48 ounces, figure 7-1.

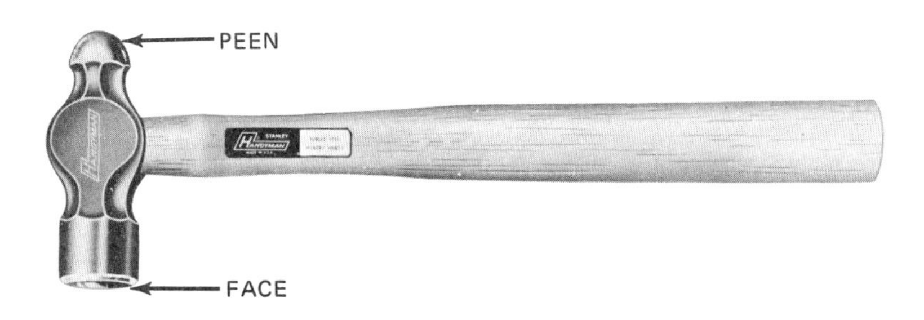

Figure 7-1 Ball peen hammer. (Courtesy of Stanley Tools.)

The Setting Hammer

One end of the head of the setting hammer has a square, flat face, and the peen end is single tapered and beveled, figure 7-2. The peen is used for peening or setting down an edge. The face is used to flatten seams. Setting hammers vary in size from 4 ounces to 20 ounces. The size used depends upon the gage of the stock and the accessibility of the work.

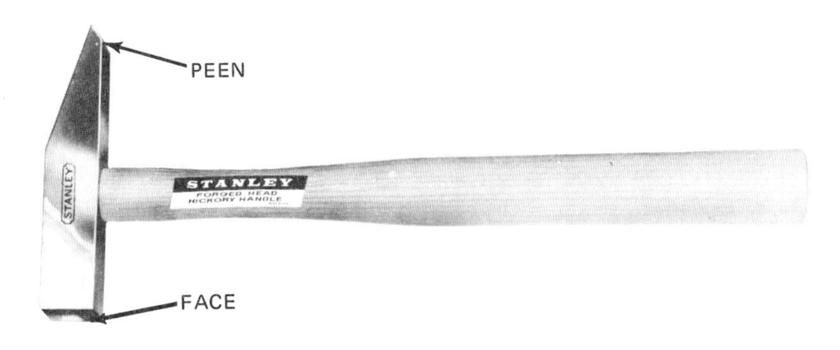

Figure 7-2 Setting hammer. (Courtesy of Stanley Tools.)

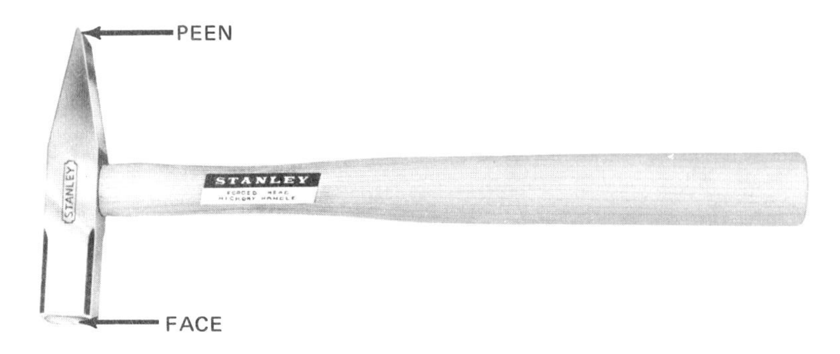

Figure 7-3 Riveting hammer.
(Courtesy of Stanley Tools.)

The Riveting Hammer

The face of the riveting hammer is slightly convex (curved), and the corners are beveled to prevent them from digging into the metal, figure 7-3. The peen end is double tapered and slightly rounded for spreading rivets. This hammer is also used for flanging and general purposes. Riveting hammers vary in sizes from 4 ounces to 30 ounces.

Combination Hammer

Although the peen end of the riveting hammer can be used for flanging, the combination hammer, sometimes called the *cross peen hammer*, is more convenient to use and must be used on jobs difficult to reach.

The combination hammer has two tapered ends at right angles to one another, figure 7-4. The peen ends must be blunt or round because a sharp peen may cut into and crack the flange before it is completely turned. This hammer is used on small jobs when it is necessary to peen by standing in front of the job.

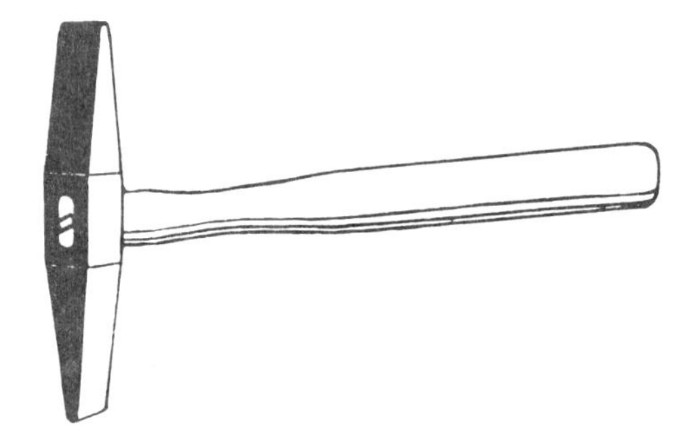

Figure 7-4 Combination hammer.

Mallets

Soft head hammers are made with soft material inserted in the faces. *Mallets*, on the other hand, have the entire head made of soft material, figure 7-5. The sheet metal worker uses wood (usually hickory), rubber, composition, and rawhide mallets. Soft head hammers and mallets yield under a blow and do not damage harder pieces of material.

The handles of hammers and mallets are made of a good quality wood stock. They are tapered at one end to prevent the heads from slipping down the handles. The head is held in place by wedges driven into the end of the handle.

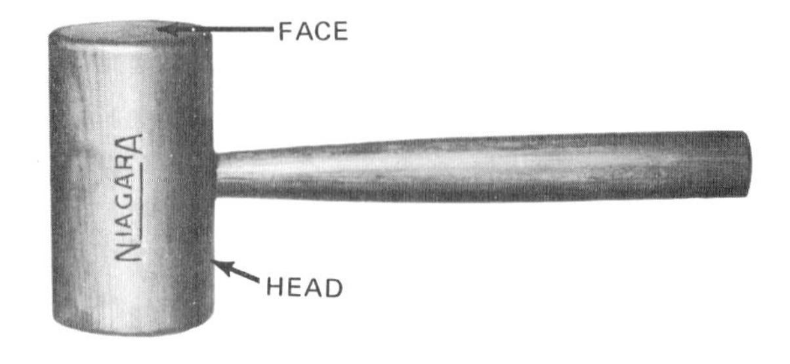

Figure 7-5 Mallet.
(Courtesy of Niagara Hand Tools)

HOW TO USE HAMMERS AND MALLETS

1. Select the proper tool for the job.
2. Inspect the hammer and mallet to make sure that the head is securely fastened to the handle, figure 7-6.

To fasten the handle, force it into the head and drive the wedge further into the handle. Split handles should be replaced.

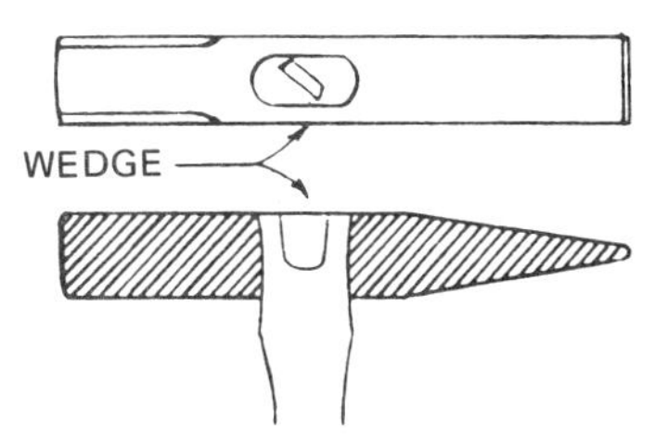

Figure 7-6 Handle securely fastened to head.

3. Grasp the tool near the end of the handle firmly but not rigidly.
4. Watch the point of action -not the head of the tool- while striking with a hammer or mallet, figure 7-7. This method assures better work. When using a mallet, strike the work with the face flat to avoid marring the surface of the stock. Use one face of the mallet for bending over sharp edges, figure 7-8, and keep the other face unmarred for flat work.

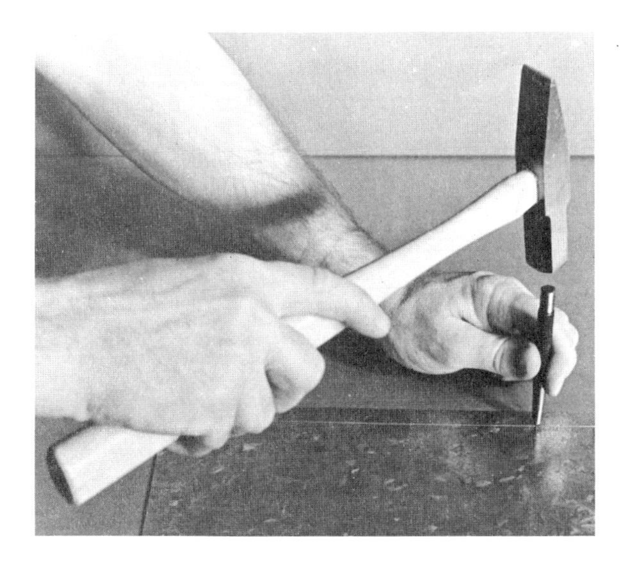

Figure 7-7 Point of action.

Figure 7-8 Flattening with a mallet.

SUMMARY REVIEW

A. Place the answers to the following questions in the column to the right.

1. List the three common hammers used by the sheet metal workers. 1. ______________________

____

____

2. Describe the face and the peen of the setting hammer. 2. ______________________

____

3. When would the sheet metal worker use a setting hammer? 3. ______________________

4. Describe the face and the peen of a riveting hammer. 4. ______________________

____

5. List the uses of the riveting hammer. 5. ______________________

____

____

6. Describe the face and the peen of a ball peen hammer. 6. ______________________

____

7. Why are mallets used on sheet metal? 7. ______________________

8. After selecting a hammer or mallet, what safety precaution should the worker observe? 8. ______________________

9. Where should a hammer or mallet be grasped? 9. ______________________

B. Insert the correct word in each of the following sentences.

1. When using a hammer, watch the point of action, not the _________ of the hammer.
2. A tool which has a head made entirely of soft material is called (a, an) ______________________.
3. If a hammer head is loose on the handle, it can be securely fastened by forcing (a, an) ______ into it.

C. Underline the correct word in each of the following sentences.

1. Hammer and mallet handles are (curved, straight, tapered) at one end to prevent the head from slipping.
2. Work should be always struck with the (edge, side, face) of a mallet to avoid marring.
3. The size of a peening hammer is designated by (length, pounds, ounces).
4. A combination hammer is used on (large, difficult-to-reach, small) jobs.

UNIT 8 *PLIERS, CLAMPS, AND VISES*

OBJECTIVES

After studying this unit, the student should be able to

- Identify and describe the various pliers used by the sheet metal worker.
- Identify and describe the types of clamping devices used by the sheet metal worker.
- Select and use the correct tool for the job.

PLIERS

Pliers commonly used by sheet metal workers are the flat nose, the side-cutting, and the combination pliers. They are used for holding and forming work and for cutting wire. Pliers are made of special steel and drop forged for added strength and resistance to brittleness.

Side-Cutting Pliers

Side-cutting pliers have flat jaws grooved to hold the work and side jaws sharpened to cut light wire, figure 8-1.

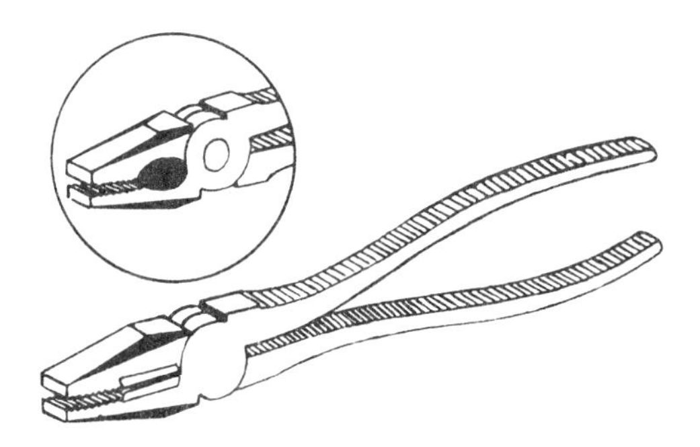

Figure 8-1 Side-cutting pliers.

Flat-Nosed Pliers

Flat-nosed pliers have flat jaws with small grooves, figure 8-2. They are used only for holding and forming work. If used carefully, these pliers do not mar the work.

Combination Pliers

Combination pliers are used for holding, cutting, and bending work, figure 8-3. The pliers are so constructed that the jaws can be adjusted for holding different sizes of work.

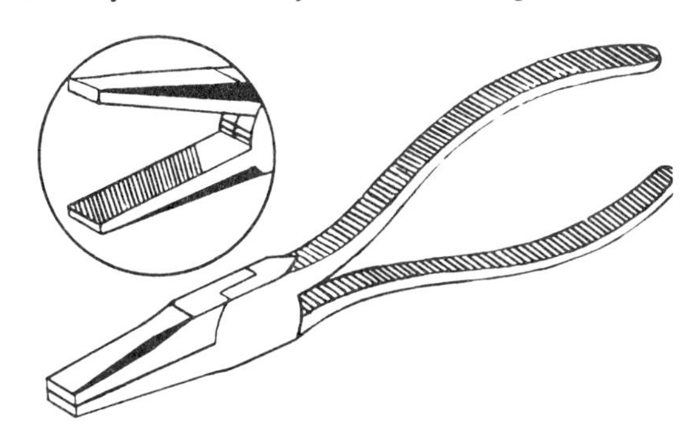

Figure 8-2 Flat nosed pliers.

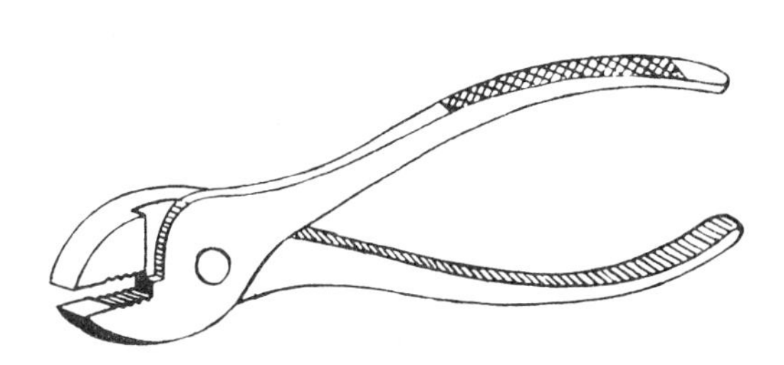

Figure 8-3 Combination pliers.

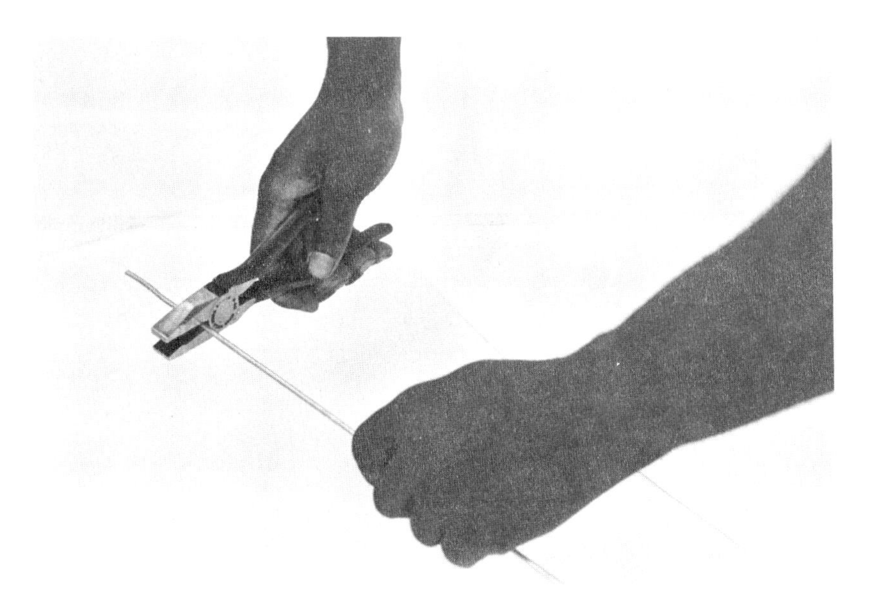

Figure 8-4
Cutting wire.

HOW TO CUT WITH PLIERS

1. Mark the line to be cut with a scriber or a piece of soapstone.
2. Grasp the piece of wire with the left hand as shown in figure 8-4.
3. Place the cutting blades of the side-cutting pliers at right angles to the wire on the cut mark with the right hand.
4. Squeeze the handles of the pliers together with the right hand, completing the cut.

HOW TO FORM WITH PLIERS

The side-cutting, flat-nosed, or combination pliers may be used to hold wire while it is being inserted in a wire fold as shown in figure 8-5. The pliers must be held at right angles to the work with a firm grip to prevent marring the work.

Figure 8-5 Holding wire with pliers.

HOW TO BEND WITH PLIERS

1. Select the proper pliers.

The pliers used depend upon the length of the lap. The flatter and wider the jaws are, the more uniform the bend.

2. Place the tip of the pliers flush with the bend line.
3. Grip the pliers firmly and bend the metal to the required angle, figure 8-6.
4. Bend the rest of the lap as in steps 2 and 3.

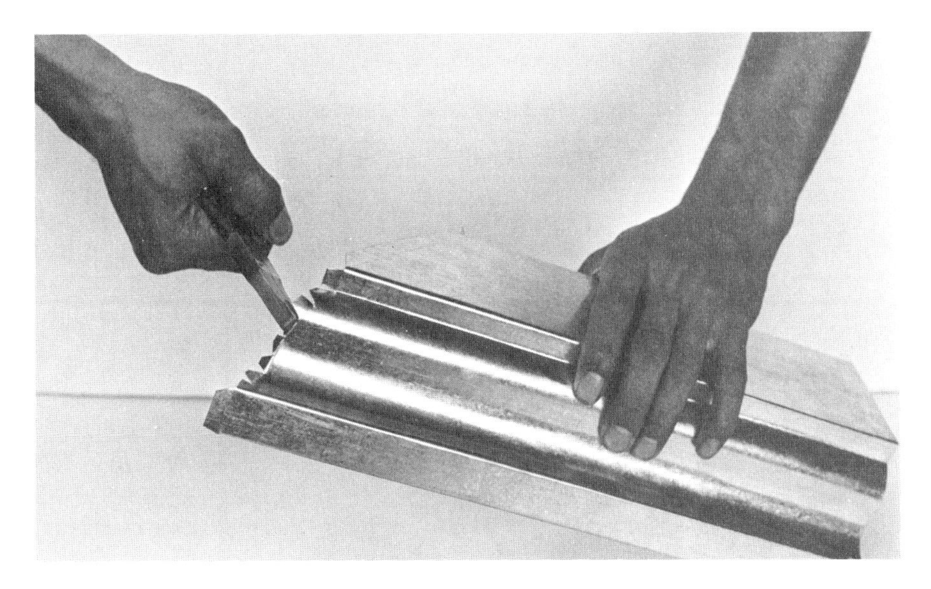

Figure 8-6 Bending with pliers.

CLAMPING DEVICES

Clamping devices used most commonly by a sheet metal worker are the C-clamp, parallel clamp, pipe bar clamp, Vise-Grips, and bench vise. These devices are used to hold and help form the work.

C-clamps

C-clamps are available in a variety of sizes and are made of malleable or cast iron.

The movable part of the C-clamp is threaded and has a swivel head. The head advances as the screw turns, but does not rotate, figure 8-7. This prevents the work from being marred.

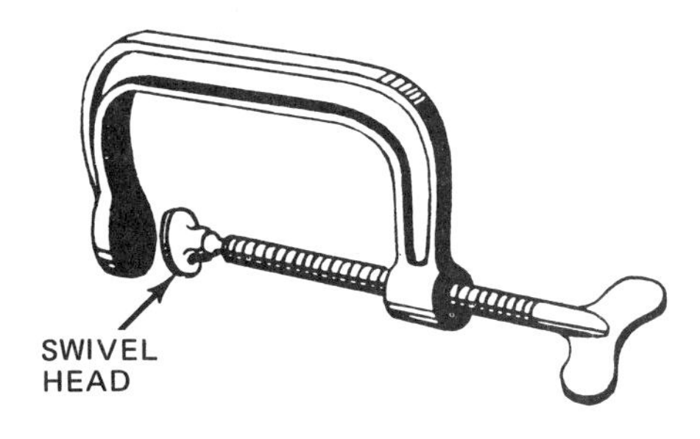

Figure 8-7 Clamp.

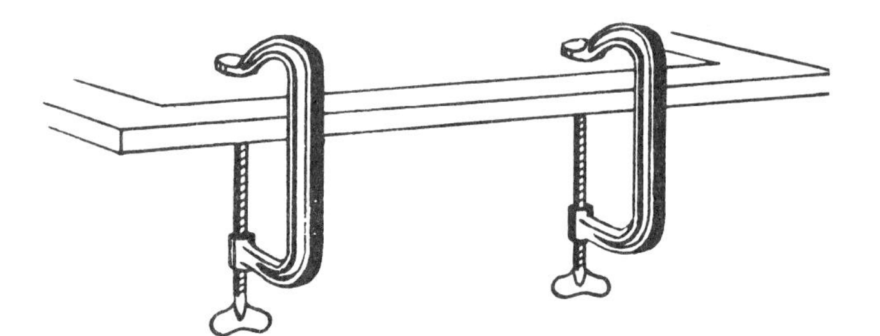

Figure 8-8 Holding work with C-clamp.

HOW TO USE C-CLAMPS

1. Clean the work and remove burrs with a file.
2. Put a drop of oil in the swivel head occasionally.
3. Place the clamps on the work with the swivel head downward, figure 8-8.
4. Tighten the clamps alternately if more than one clamp is being used.

Parallel Clamps

The parallel clamp has two knurled screws. The screw near the work tends to hold the jaws together. The screw on the outside tends to push the jaws apart. The resulting leverage clamps the work when the jaws are parallel, figure 8-9A. To get good results, the jaws must be parallel so that the full surface of the jaws covers the work. If the jaws do not clamp the work evenly, the work will slip, figure 8-9B.

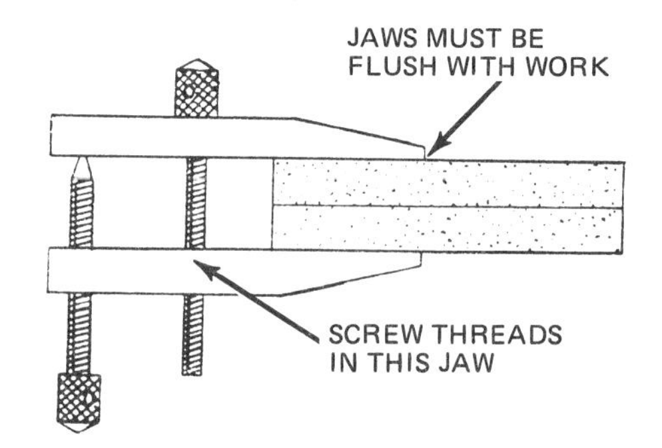

Figure 8-9a Work correctly clamped.

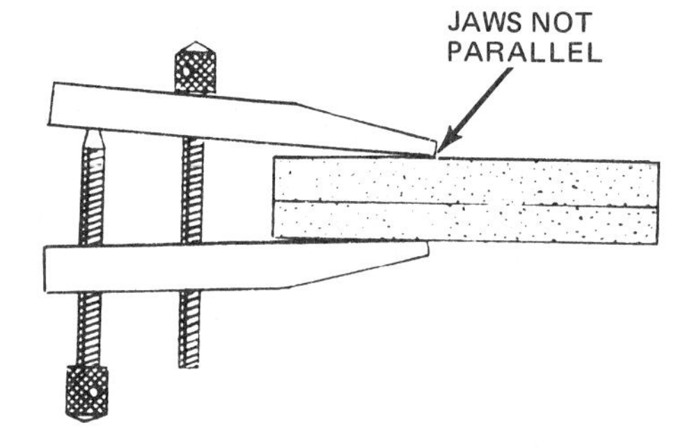

Figure 8-9b Work incorrectly clamped.

HOW TO USE PARALLEL CLAMPS

1. Clean the surfaces to be clamped and remove burrs with a file.
2. Open the clamps by turning the screws. Place the clamps so that as much of the jaws are in contact with the work as possible.
3. Adjust the clamps until they are open a slight amount, figure 8-10A.
4. Turn the rear screw until the jaws are closed. When the jaws are correctly closed, they rest flat on the work, figure 8-10B.

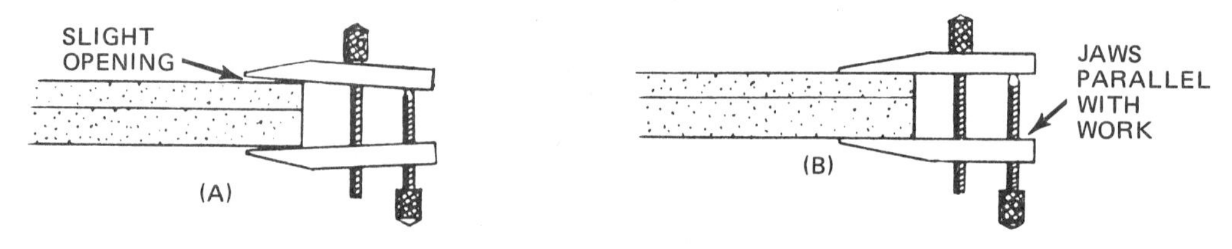

Figure 8-10 Adjusting a parallel clamp.

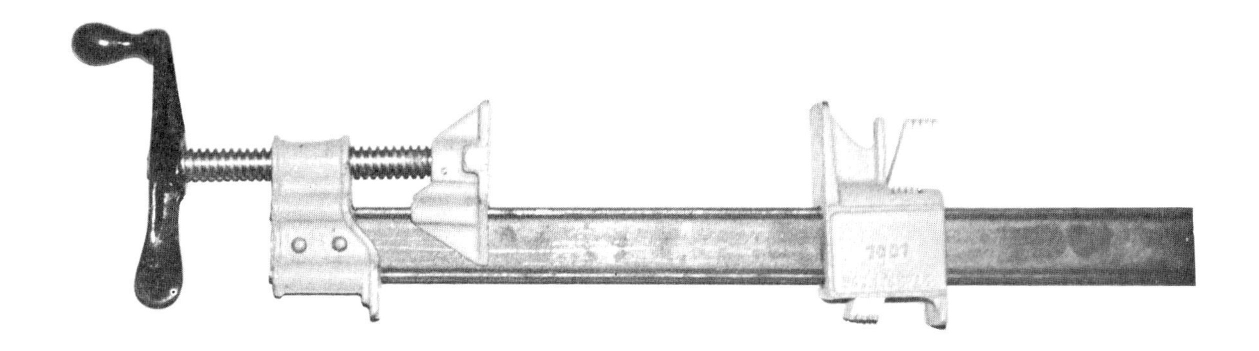

Figure 8-11 Bar clamp.

Bar Clamps

Bar clamps, figure 8-11, are available in a variety of designs and sizes depending on the shape of the bar. Bars can be square, rectangular, round, or I-shaped. Clamps are made to fit the bar size and shape. When pipe is used, clamps for standard 1/2-inch and 3/4-inch diameter pipe are available. Figure 8-12 shows the most commonly used style with a 3/4-inch diameter pipe as the bar. Any length can be used. A foot is screwed on one end of the pipe, while the head slips on the other end, automatically gripping the bar at any point. A multiple disc clutch permits instant adjustment and securing of the foot to fit the work. If a spreading action is required, both the head and foot are reversed, figure 8-13. On some clamps only the foot is reversible.

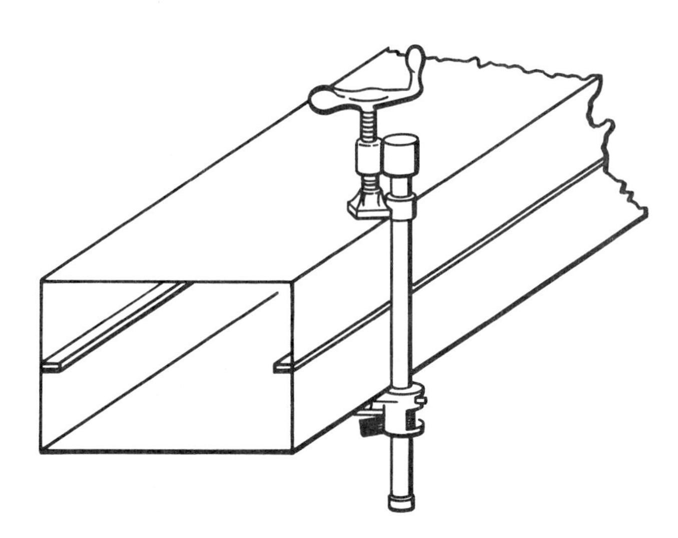

Figure 8-12 Bar clamp used in duct fabrication.

Figure 8-13
Bar clamp reversed.

HOW TO USE A BAR CLAMP

1. Select a bar clamp in which the length of the bar is nearest the size of the work being clamped.
2. Place the head and foot on each side of the work and slip the head up to the work.
3. Turn the handle on the foot clockwise to tighten or counterclockwise to loosen.
4. If more than one clamp is used, tighten the clamps alternately.

Vise Grip Clamps

The Vise-Grip® wrench is a general purpose wrench used for holding or turning. The Vise Grip sheet metal clamp is used to hold sheet metal pieces together during fabrication. Vise Grip C-clamps are used to hold sheet metal stock and patterns firmly in place on the bench for layout. As the name implies, Vise Grip welding clamps are used to hold pieces together while being welded.

HOW TO USE A VISE GRIP CLAMP

1. Select the proper type of Vise Grip for the job.
2. Adjust the jaw opening to fit the job with the handles closed.
3. Open the handles and turn the adjustment screw 1/4 turn tighter.
4. Place the jaws on job and squeeze the handles into the lock position.
5. To open the clamp, press the release lever.

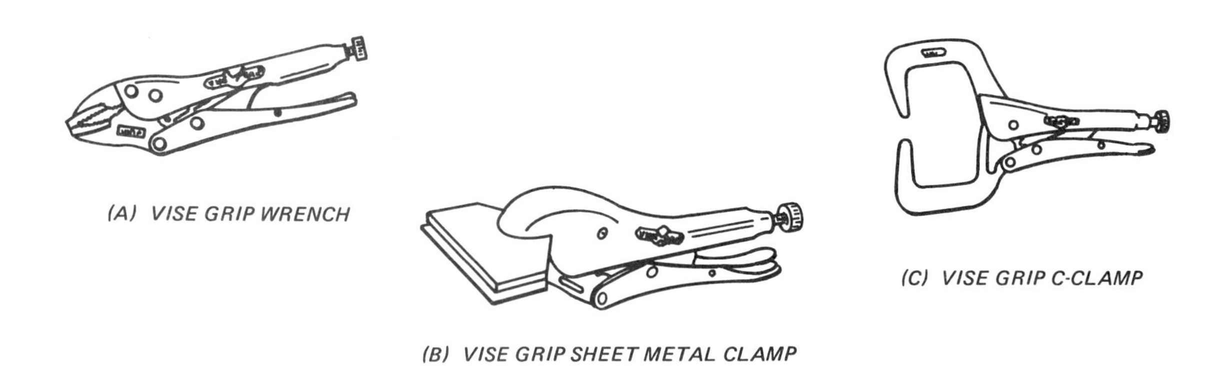

(A) VISE GRIP WRENCH

(B) VISE GRIP SHEET METAL CLAMP

(C) VISE GRIP C-CLAMP

Figure 8-14

The Bench Vise

The bench vise holds the work between its jaws. The handle turns the clamp screw which moves the outside jaw to clamp the work, figure 8-15. On many vises, the immovable jaw is fixed to a swivel base that may be rotated on its axis to secure the work in the proper position if it must be held at an angle to the normal position of the vise.

The vise jaws are hardened and serrated to grip the work rigidly. Work that must not be marred can be protected by using soft removable jaws, figure 8-16. The removable jaws are made of soft materials such as brass, copper, and leather. These materials do not mar the surfaces, but the jaws still grip the work securely.

Figure 8-15 Bench vise.

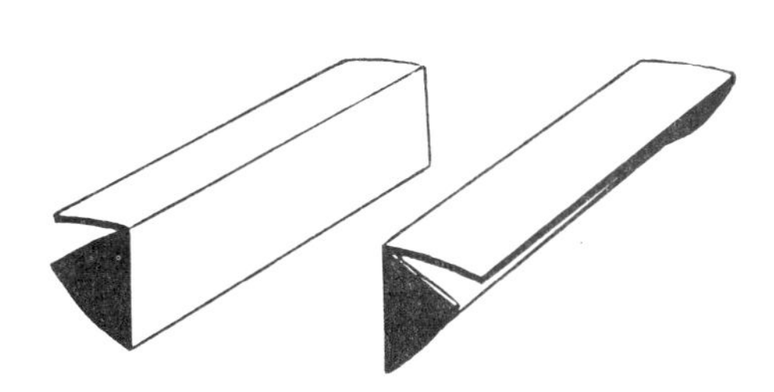

Figure 8-16 Soft removable jaws.

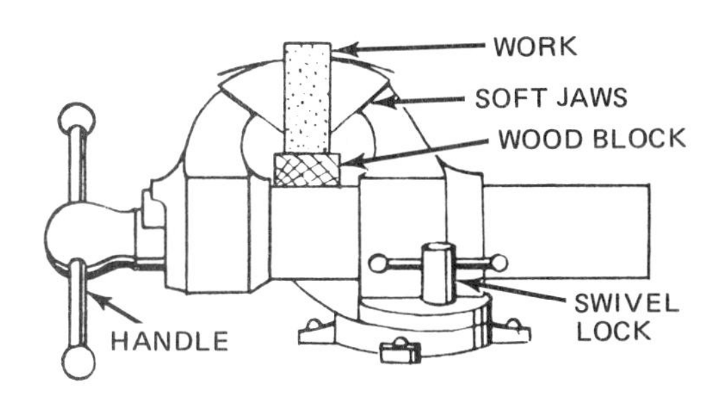

Figure 8-17 Bench vise.

HOW TO USE THE BENCH VISE

1. Clean the vise frequently--at least daily. Oil the clamp screw occasionally.
2. Place a pair of soft jaws over the jaws of the vise if finished work is being clamped. Put a block below the work to prevent it from slipping down in the vise, figure 8-17.
3. Tighten the work by rotating the handle. Be careful not to crack thin sections or force circular pieces out of round.

CAUTION It is poor practice to tighten the handle of the vise by striking it with a hammer.

SUMMARY REVIEW

A. Place the answers to the following questions in the column to the right.

1. What advantage do combination pliers have for holding various sizes of work? 1. ____________________
2. A wire edge of 16-gage (.0625) wire is needed on a wastebasket. Which types of pliers should be used for each operation? 2. ____________________

3. In backing out a series of screws, pliers must be used to finish the job. What type should be selected? 3. ______________

4. What is the best type of pliers for holding irregular objects? 4. ______________

5. Name the two necessary parts of any clamping device. 5. ______________

6. In using a parallel clamp, how must the jaws be placed to get maximum holding power? 6. ______________

7. List the three parts of a bar clamp. 7. ______________

8. Identify four types of Vise Grips used by the sheet metal worker. 8. ______________

9. How are bench vise jaws designed to grip the work rigidly? 9. ______________

10. What is the purpose of removable, soft vise jaws? 10. ______________

B. Insert the correct word in each of the following sentences.

1. The main purpose of pliers is to ________ a piece of stock.
2. If flat-nosed pliers are used properly, they will not ______ the material.
3. The purpose of the ________ head on a C-clamp screw is to prevent____________.
4. The bench vise is used for ________work between its jaws.
5. A vise should not be tightened by _______ the handle with ____________ .

C. Underline the correct word in each of the following sentences.

1. In order to drill a series of holes in 12 pieces of 16-gage black iron, 12 by 14 inches, they should be held together with (a bench vise, bar clamps, parallel clamps).
2. A large duct should be held together with (a C-clamp, bench vise, bar clamp).
3. If the flanges on a pipe of 14-gage black iron are to be welded, they should be held in position with (a bench vise, bar clamp, welding clamps).
4. If a small piece of 26-gage galvanized iron is to be drilled in a drill press, it should be held with (the hand, bench vise, vise grip wrench).
5. A sheet of 26-gage galvanized iron can be clamped on a flat surface with (bar clamps, parallel clamps, a bench vise).

UNIT 9 WRENCHES

OBJECTIVES

After studying this unit, the student should be able to

- Identify and describe wrenches used by the sheet metal worker.
- Select the proper wrench for the job.
- Use each wrench correctly.

A great many wrenches have been designed to meet the needs of the sheet metal worker. Although special wrenches are made for particular purposes, wrenches generally fall into two classifications, adjustable and stationary. The adjustable types are those that fit numerous sizes of fastening devices, and the stationary types are made to be used on only one size fastener.

ADJUSTABLE-TYPE WRENCHES

The Monkey Wrench

The monkey wrench, figure 9-1, is an all-purpose wrench designed specifically for turning large octagonal and square heads of bolts and nuts.

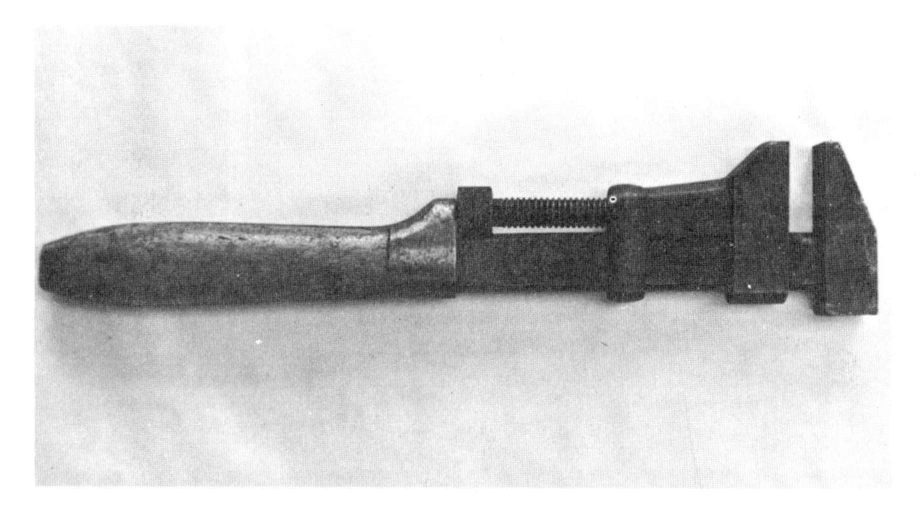

Figure 9-1 Monkey wrench.

The Adjustable Wrench

Adjustable wrenches, figure 9-2, are available in a variety of sizes from 4 inches to 18 inches. The jaws of an adjustable wrench are turned at an offset from the handle to facilitate working in tight places. With this offset, the wrench can be rotated one-quarter turn, and then reversed to make another quarter turn in the same small opening.

Figure 9-2 Adjustable wrench. (Courtesy of Stanley Tools.)

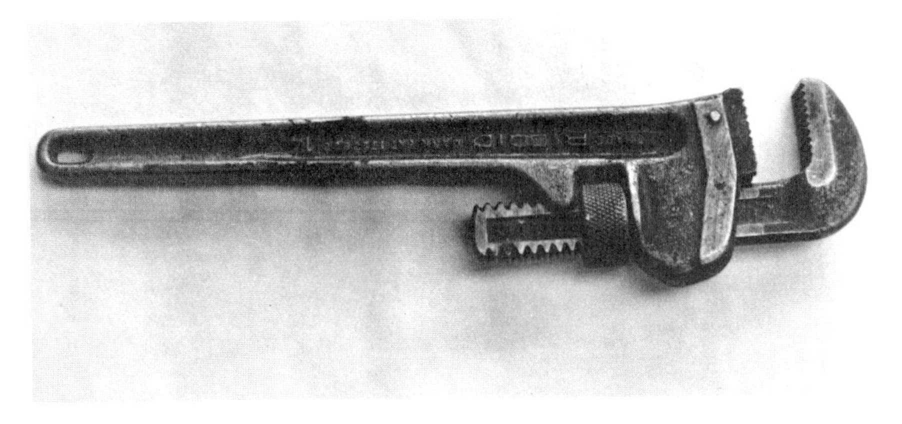

Figure 9-3 Pipe wrench.

The Pipe Wrench

The pipe wrench, figure 9-3, is used for turning pipes and other round objects. Heavy teeth are cut in the jaws to grip the pipe or rod. A pipe wrench grips only when pressure is applied to the handle toward the face side of the jaws. A pipe wrench should never be used on a nut or a bolt head unless the corners of the fastener have been rounded so that no other type of wrench can grip it.

HOW TO USE AN ADJUSTABLE-TYPE WRENCH

1. Place the wrench on the part to be moved with the jaws facing in the direction in which the work is to be turned, figure 9-4.
2. Adjust the jaws until they fit tightly. If the nut or bolt does not start easily, it may be loosened by striking the handle of the wrench with the heel of the hand.

Too much leverage on the wrench breaks off small fastenings. Use a wrench of the correct length for the job.

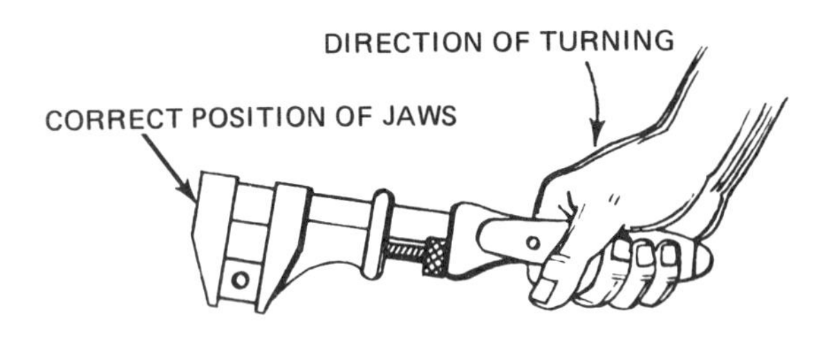

Figure 9-4 Using the monkey wrench.

STATIONARY WRENCHES

The Open-End Wrench

The open-end wrench, figure 9-5, is a nonadjustable wrench made of forged, high grade steel. An opening on each end provides for two different sizes of bolts or nuts. The jaws are sometimes offset to allow efficient use in close places. The size of the opening between the jaws determines the size of the wrench. The size is stamped on each end to help the worker.

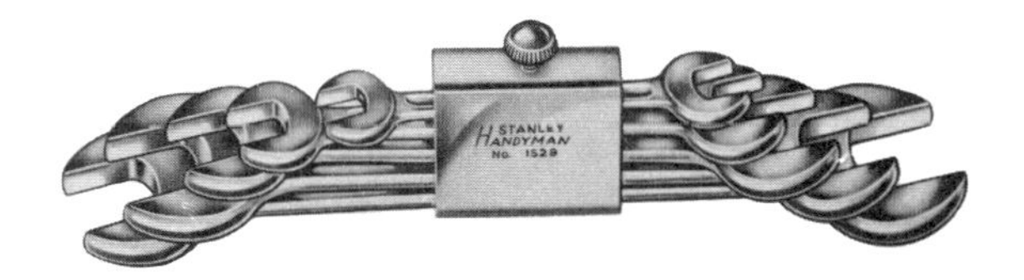

Figure 9-5 Open-end wrench set.
(Courtesy of Stanley Tools.)

All types of open-end wrenches are available in different sizes. They are sold usually in sets of twelve, in sizes ranging from 5/16 inch to 1 1/2 inches.

Before using an open-end wrench, make certain that it fits the nut or bolt snugly. Always pull on the wrench to move it; never push on a wrench.

The Box Wrench

The box wrench, figure 9-6, has two ends which are offset. Unlike the open-end wrench, the openings are enclosed. The inside of the opening is divided into twelve grooves which make it possible to completely rotate hex nuts where the swing of the wrench is limited. Box wrenches are usually sold in sets of twelve in sizes ranging from 5/16 inch to 1 1/2 inches.

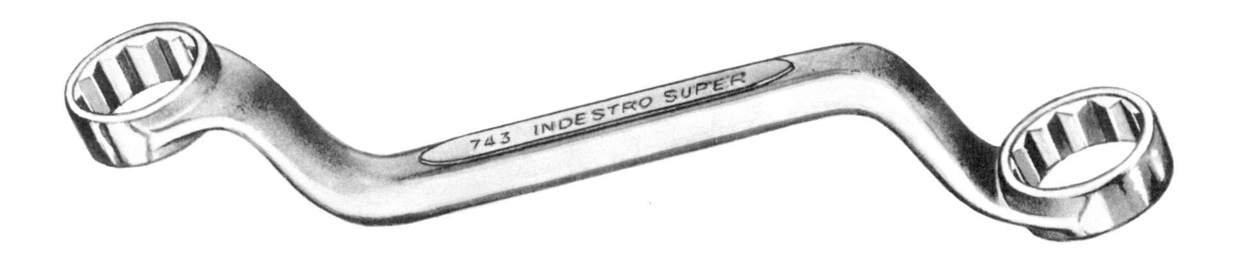

Figure 9-6 Short handle box wrench. (Courtesy of Indestro Manufacturing Co.)

The Socket Wrench

Socket wrenches, figure 9-7, come in sets or kits, which consist of a ratchet handle and a number of forged-steel sockets in a series of sizes. The total range of sizes available is 1/4 inch to 2 1/2 inches. There are two holes in each socket: one is a square hole to fit the ratchet handle, and the other is a round hole with grooves like the box wrench that slips over the nut or bolt head. Socket wrenches have the advantages of speed and positive action of rotation in either direction.

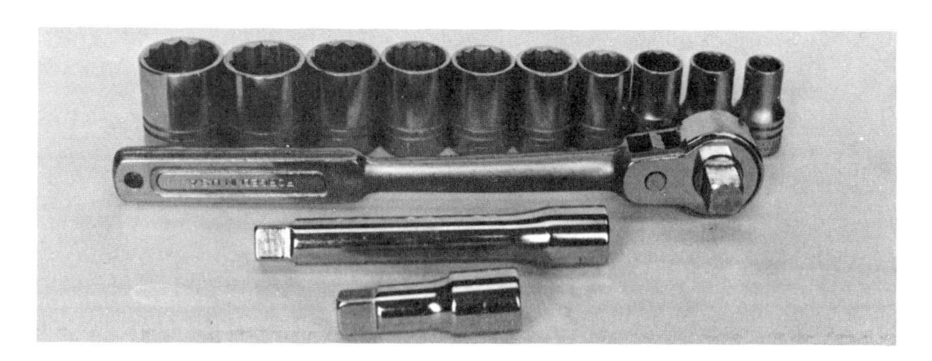

Figure 9-7
Socket wrench set.

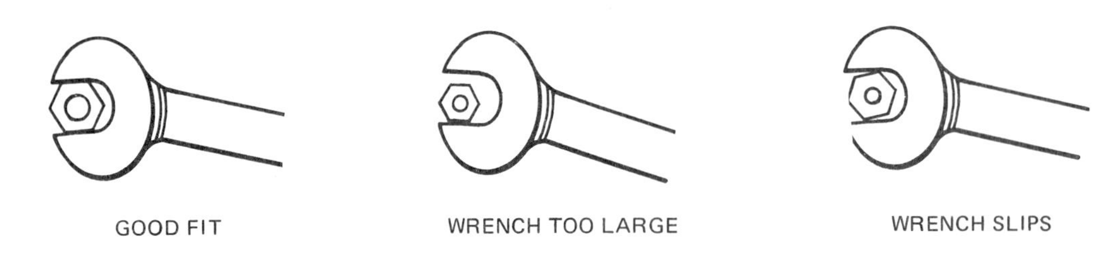

Figure 9-8

HOW TO USE A STATIONARY-TYPE WRENCH

When using a stationary wrench, it is important to choose a size which fits the work snugly, figure 9-8. If the wrench slips on the work because of a loose fit between the work and the jaws, the worker may injure his hands and damage the bolt head.

Allen Wrenches

Allen wrenches, figure 9-9, are L-shaped pieces of hexagonal tool steel. They are made to fit the hexagonal socket in Allen headless setscrews. These setscrews are used to fasten pulleys or collars in place where screws or bolts with protruding heads cannot be used.

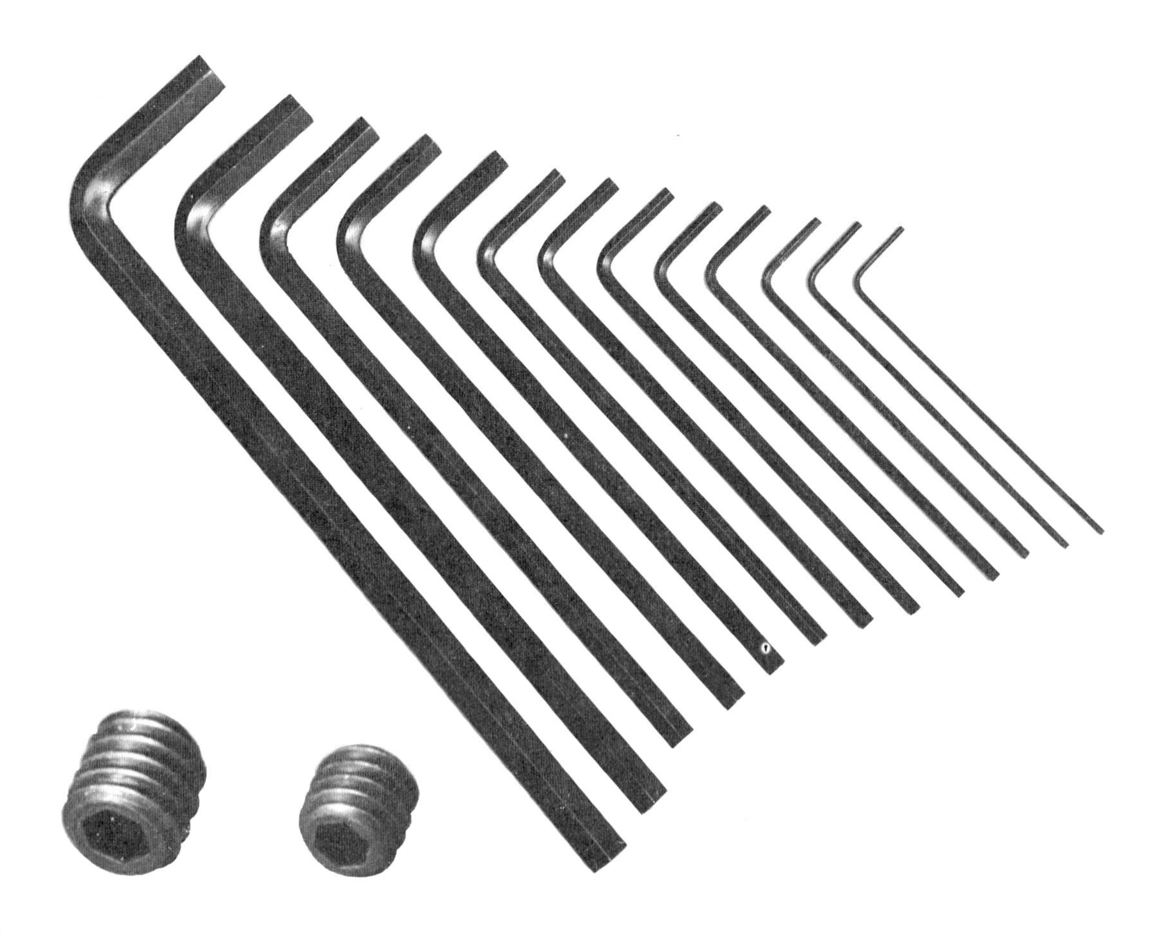

Figure 9-9 Set of Allen wrenches and headless setscrews.

HOW TO USE AN ALLEN WRENCH

1. Insert the short end of the wrench into the head of the Allen screw.
2. Apply force on the long end of the wrench to tighten or loosen the screw.

Select a wrench to fit the screw exactly. After loosening the screw with the short end of the wrench, the screw can be removed faster by inserting the long end of the wrench into the screw head.

SUMMARY REVIEW

A. Place the answers to the following questions in the column to the right.

1. List the two general types of wrenches in use. — 1. ______ ______
2. When would the worker select a monkey wrench for a job? — 2. ______
3. Why are the jaws of an open-end wrench sometimes offset from the handle? — 3. ______
4. What is the smallest amount of turning that can be accomplished with an adjustable wrench? — 4. ______
5. What type of objects are turned with a pipe wrench? — 5. ______
6. Why are the jaws of a pipe wrench grooved? — 6. ______
7. List four steps that the worker follows in using an adjustable wrench. — 7. ______ ______ ______ ______
8. List three types of stationary wrenches in use. — 8. ______ ______ ______
9. How does a box wrench differ from an open-end wrench? — 9. ______
10. What advantage does a box wrench have over an open-end wrench? — 10. ______
11. List two reasons for using a socket wrench in tightening or loosening a bolt. — 11. ______ ______
12. Select the wrench to be used for each of the objects shown below. — 12. ______ ______ ______ ______

(A) (B) (C) (D)

B. Insert the correct word in each of the following sentences.

1. When tightening or loosening a nut or a bolt with a wrench, it is important to have a ________ fit and apply leverage ________ the body.

2. Leverage should never be applied to a wrench by ________ on the handle instead of __________.

C. Underline the correct word in each of the following sentences.

1. A pipe wrench should be used to turn (nuts, bolts, round objects).

2. To tighten the octagonal nut on a 3/4-inch diameter bolt, (a pipe wrench, monkey wrench, socket wrench, adjustable wrench) is the best tool to use.

3. In tightening a bolt where the swing is limited, (an open-end wrench, box wrench, adjustable wrench) should be used.

UNIT 10 SQUARES

OBJECTIVES

After studying this unit, the student should be able to

- Identify and describe the types and parts of a square.
- List the uses of a square.
- Select and use a square properly for a job.

The steel square and combination square are used by the sheet metal worker to lay out 90-degree and 45-degree angles. In addition, measurements can be made, and the straightness or squareness of a surface or an edge can be checked with these tools.

The Steel Square

The steel square, figure 10-1, has 1/16- or 1/8-inch graduations on the inside and outside edges, and it lies flat on the surface to be measured. The long part of the square is called the *body* or *blade*; and the short part, the *tongue*. The outside corner is termed the *heel*. Steel squares are available in 12-, 18-, and 24-inch blade lengths.

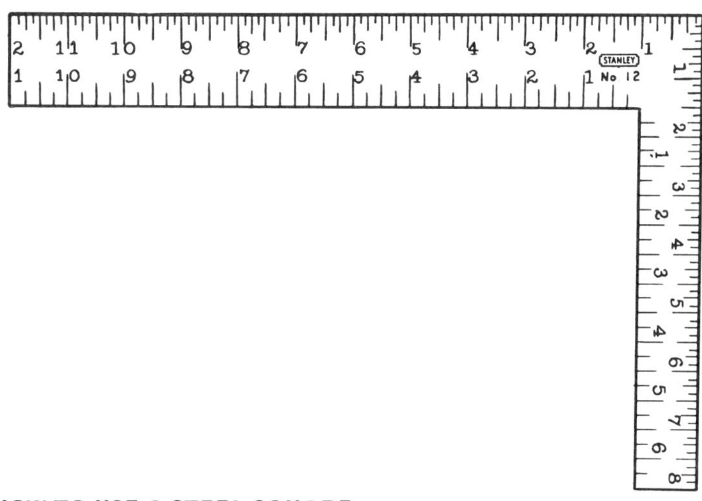

Figure 10-1 Steel square.
(Courtesy of Stanley Tools)

HOW TO USE A STEEL SQUARE

A. *90-Degree Angle or Perpendicular Line*

1. Lay out the base line.
2. Place the blade edge of the square on the base line.
3. Adjust the heel to the desired point for the perpendicular line.
4. Scribe the line along the tongue edge.

B. *Square Edge*

1. Place the inner edge of the blade against the straight edge of the stock.
2. Scribe the line along the edge of the tongue.

C. *Angle Layout*

1. Scribe a horizontal base line on the surface of the stock.
2. Place the tongue of the square on the base line.

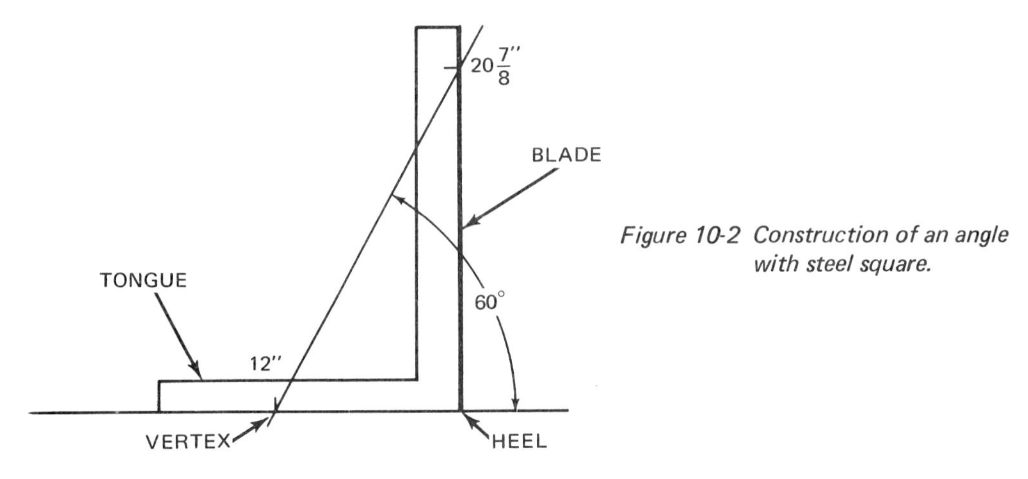

Figure 10-2 Construction of an angle with steel square.

3. Adjust the 12-inch mark on the tongue to the vertex of the desired angle, figure 10-2.
4. Mark the vertical distance, table 10-1, for the desired angle along the blade edge of the square.
5. Scribe a line to connect the points for the completed angle.

Angle	Tongue	Blade
15°	12"	3 7/32"
22½°	12"	4 31/32"
30°	12"	6 15/16"
45°	12"	12"
60°	12"	20 7/8"

TABLE 10-1 ANGLE LAYOUT

THE COMBINATION SQUARE

The combination square has adjustable heads which may be clamped to the blade at any desired distance from the end of the blade, figure 10-3.

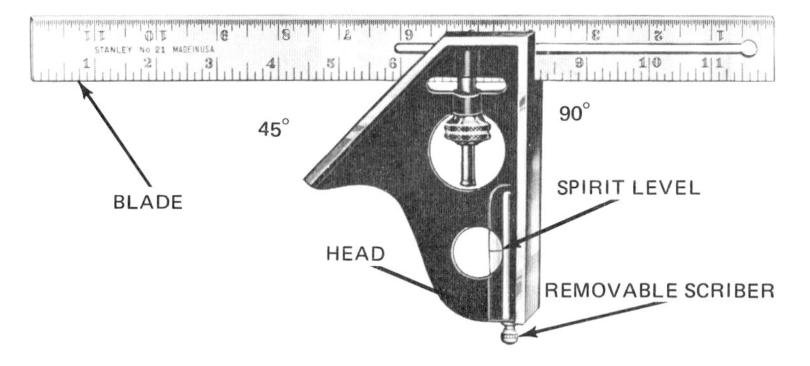

Figure 10-3 Combination square. (Courtesy of Stanley Tools)

The *square head* is made of cast iron with machined edges of 90 degrees and 45 degrees from the edge of the blade line. The *blade* is made of tempered steel which has been ground and polished. It is stamped in graduations of eighths, sixteenths, and thirty-seconds of an inch. The blade can be removed from the head and used as a short straightedge. The head is provided with a spirit level and a removable scriber as shown in figure 10-3.

A combination square may be used as an inside or outside try square, plumb and level, depth gage, or marking gage.

Besides the square head, the center head and protractor head are used with the blade of the combination square. The center head is used to locate the center of round stock, and the protractor head is for laying out angles up to 180 degrees.

HOW TO USE THE COMBINATION SQUARE

A. *As a Straightedge*

1. Remove the blade from the head by depressing the thumbscrew and pulling the blade from the head.
2. Place the blade on a surface and scribe a line along the edge as shown in figure 10-4.

Figure 10-4 Using the combination square as a straightedge.

B. *For Gaging*

1. Clamp the blade to the head set at the desired width.
2. Hold a scratch awl at the end of the blade and move the combination square and scratch awl along the sheet, making the necessary gage mark, figure 10-5.

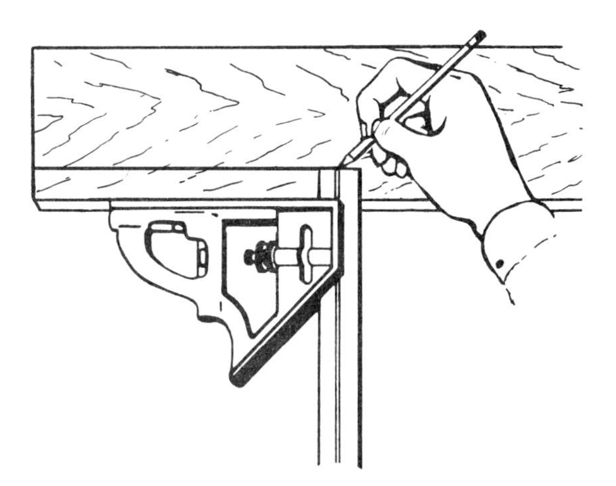

Figure 10-5 Using the combination square as a gage.

SUMMARY REVIEW

A. Place the answers to the following questions in the column to the right.

1. Name two types of squares common in sheet metal work. 1. ______________________ ______________________
2. List the parts of a steel square. 2. ______________________ ______________________
3. Where is the heel on a steel square? 3. ______________________
4. List steps necessary to square up the rough edge of the duct shown below from point A. 4. ______________________ ______________________ ______________________ ______________________

(A)

5. Give the three common sizes of steel squares. 5. ______________________
6. List four steps necessary in laying out a 30-degree angle with a steel square from a given base line and vertex point. 6. ______________________ ______________________ ______________________ ______________________
7. Identify the four possible parts of a combination square set. 7. ______________________ ______________________ ______________________ ______________________
8. State the three uses of a combination square other than as a try square. 8. ______________________ ______________________ ______________________
9. What is the purpose of the centering head in a combination square set? 9. ______________________
10. How is the blade removed from the square head of a combination square? 10. ______________________

B. Insert the correct word in each of the following sentences.

1. The steel square is made to lay ______ on a surface.
2. The square head of a combination square is provided with ____________ and ____________ .
3. The head of a combination square is ____________ on the blade.

C. Underline the correct word in each of the following sentences.

1. A combination square can be used as (a screwdriver, pry, depth gage).
2. The squareness of the edge of a piece of stock can be easily checked with (a rule, tape, steel square).
3. The protractor head on the combination square is used to lay out (straight lines, angles, curves).

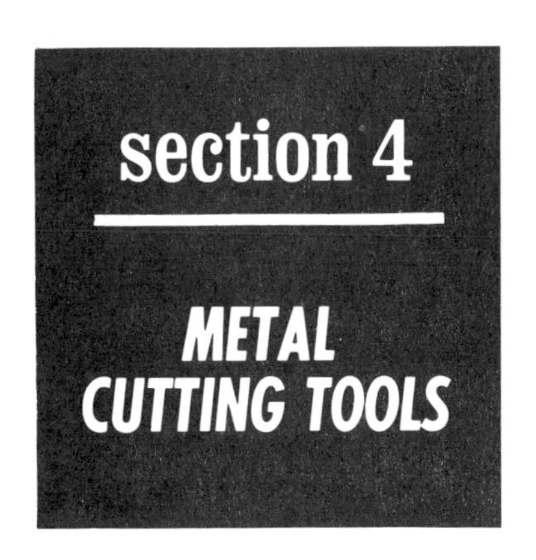

UNIT 11 *HAND SNIPS*

OBJECTIVES

After studying this unit, the student should be able to

- Identify and describe the types of hand snips used in the sheet metal trade.
- Select the right snips for the job.
- Use and care for each type of snips correctly.

Various kinds of hand snips are used for cutting and notching sheet metal. Hand snips are necessary because the shape, construction, location, and position of the work to be cut often prevent the use of machine cutting tools.

Snips should be selected so that they cut easily and are not tiresome to use. They should, therefore, have sharp cutting edges, have the joints centered for good leverage, be well balanced, and have the bows made to fit the hand.

Hand snips are roughly divided into two groups: those for straight cuts, which are the straight snips, combination snips, bulldog snips, and compound lever shears; and those for circular cuts, which are the circle snips, hawk's-bill snips, trojan snips, and aviation snips. There are many additional kinds of snips available for special purposes.

Different kinds of snips require different techniques for their correct use. Combination and aviation snips are more practical than the straight or circle snips because of the greater variety of cutting which can be done with them.

Snips are made of forged steel. Special high power snips with alloy steel blades, branded "special for alloy and stainless steels" are used to cut stainless steel and other hard materials. These high power snips are available in the straight, combination, and bulldog styles. The snips for heavy gages and hard materials have longer handles and, in some cases, special arrangements of levers to make cutting easier.

Compound lever shears, bench shears, and Beverly shears are fastened to the bench, have long handles, and are used to cut materials 1/16 inch thick and heavier. They are being replaced by squaring shears and electric shears in most shops.

Wire should not be cut with snips, and metal should not be cut by hammering on the handles or jaws of the snips.

STRAIGHT-CUTTING SNIPS

Straight Snips

Straight snips have straight jaws for straight line cutting. To maintain strength, the ends should not be unduly pointed, figure 11-1.

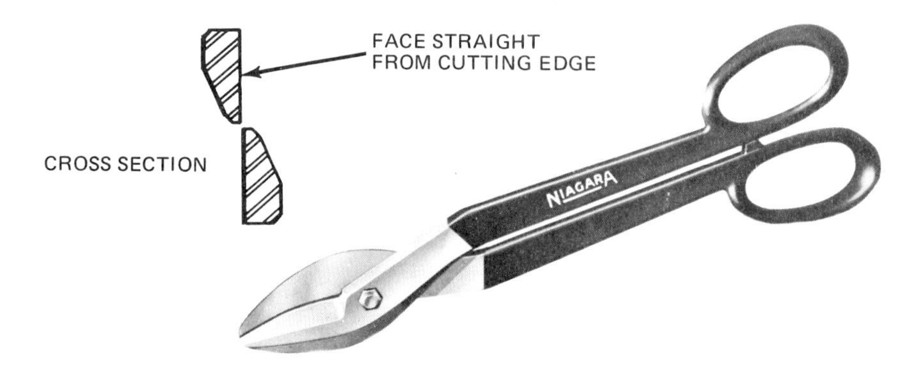

Figure 11-1 Straight snips. (Courtesy of Niagara Hand Tools.)

Straight snips are supplied in many different sizes. The jaws or cutting edges vary in length from 2 to 4 1/2 inches. The overall length of these snips vary from 7 to 15 3/4 inches. The various sizes of straight snips cut different thicknesses of mild steel; 20-gage steel is the maximum thickness for the larger snips. These snips are available for right- and left-hand use.

Combination Snips

Combination snips have straight jaws for straight cutting, but the inner faces of the jaws are sloped for cutting curves and irregular outlines, figure 11-2. These snips are available in the same sizes and capacities as straight snips.

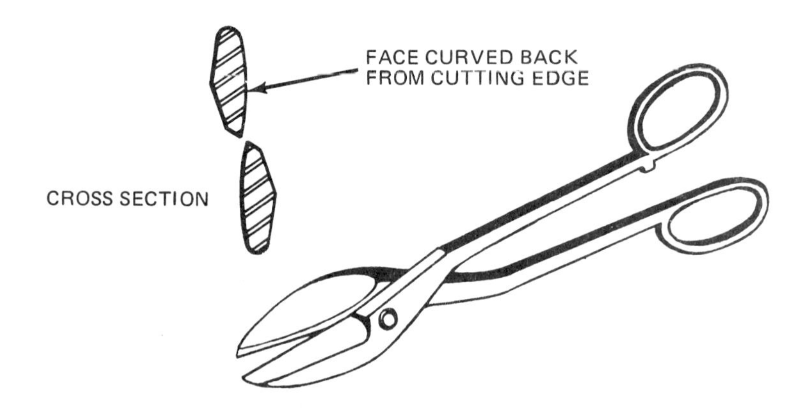

Figure 11-2 Combination snips.

Bulldog Snips

Bulldog snips are like combination snips, but they have shorter cutting blades with longer handles for greater leverage, figure 11-3. The blades are inlaid with a special alloy steel for cutting stainless steel. Bulldog snips can cut 16-gage mild steel. The blades are 2 1/2 inches long, and the overall length of the snips varies from 14 to 17 inches.

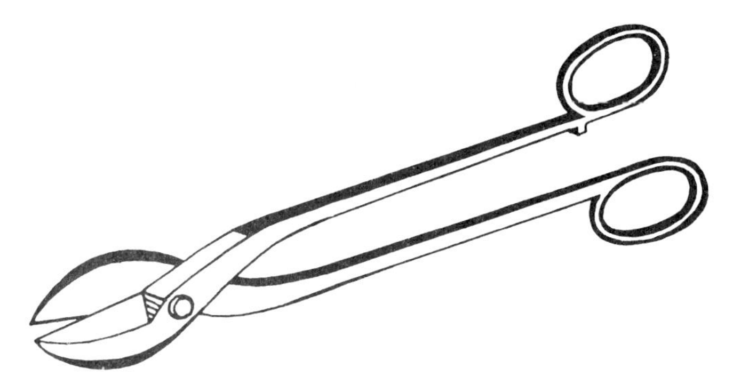

Figure 11-3 Bulldog snips.

HOW TO USE STRAIGHT, COMBINATION, AND BULLDOG SNIPS

A. *Straight Cuts*

1. Select the proper snips. For mild steel, 22 gage and lighter, use either straight or combination snips. For 16 to 20 gage mild steel, use bulldog snips.
2. Inspect the snips to see that they are properly adjusted. If adjustment is necessary, refer to "How to Oil and Adjust Snips," page 61.
3. Grasp the snips in the right hand and the narrowest part of the sheet in the left hand, figure 11-4. Rest the snips and sheet on the bench if necessary.

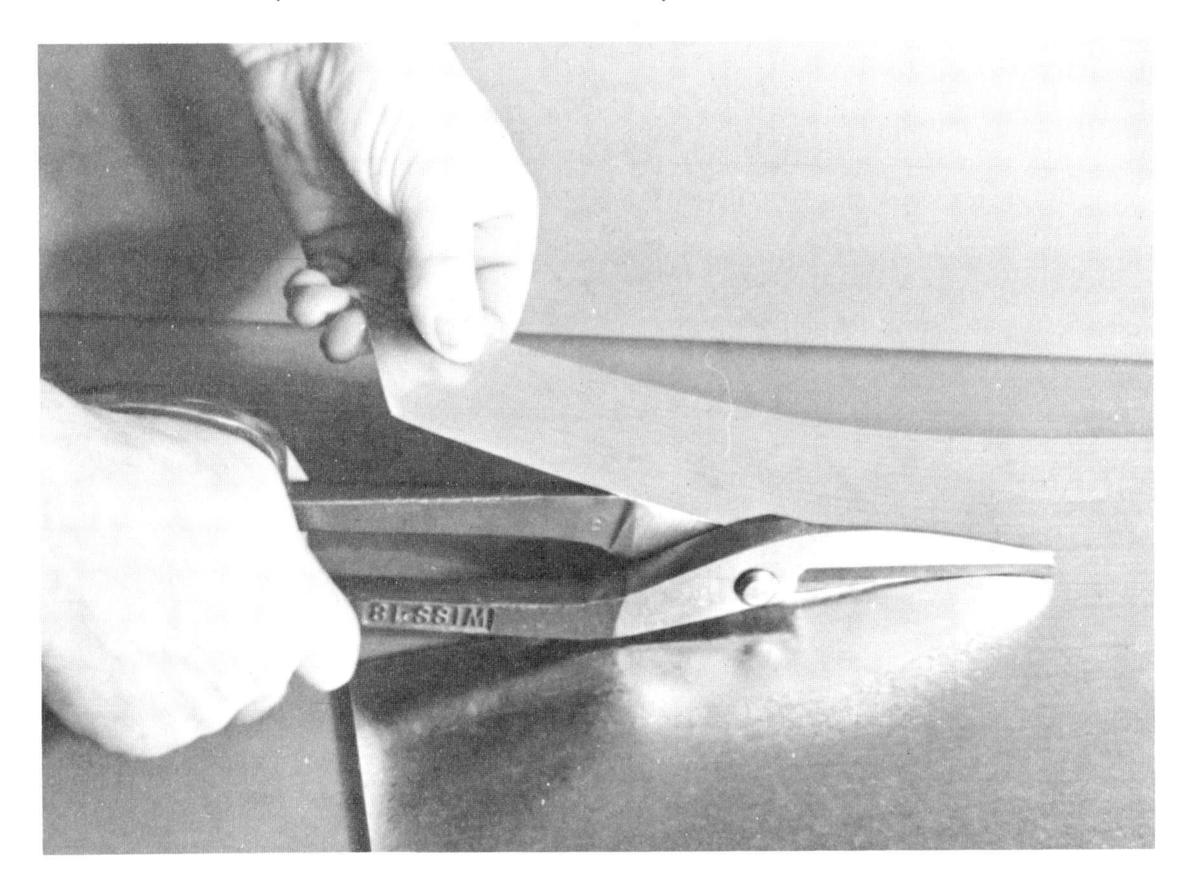

Figure 11-4 Making a straight cut.

4. Open the blades of the snips as far as possible and start the cut at the edge of the sheet. The snips should always be held at right angles to the sheet to be cut.
5. Cut the sheet by closing the blades just short of the full length to prevent leaving jagged edges, figure 11-4. The length of each cut is determined to a certain extent by the gage of the sheet: the heavier the metal, the shorter the cut.
6. Start the snips at the extreme end of the preceding cut.
7. Finish the cut, keeping the snips on the line by changing the direction of the snips if necessary.

File off any jagged edges or slivers after cutting.

B. *Outside Curved Cuts*

1. Cut off the corners of the metal to facilitate handling, figure 11-5.

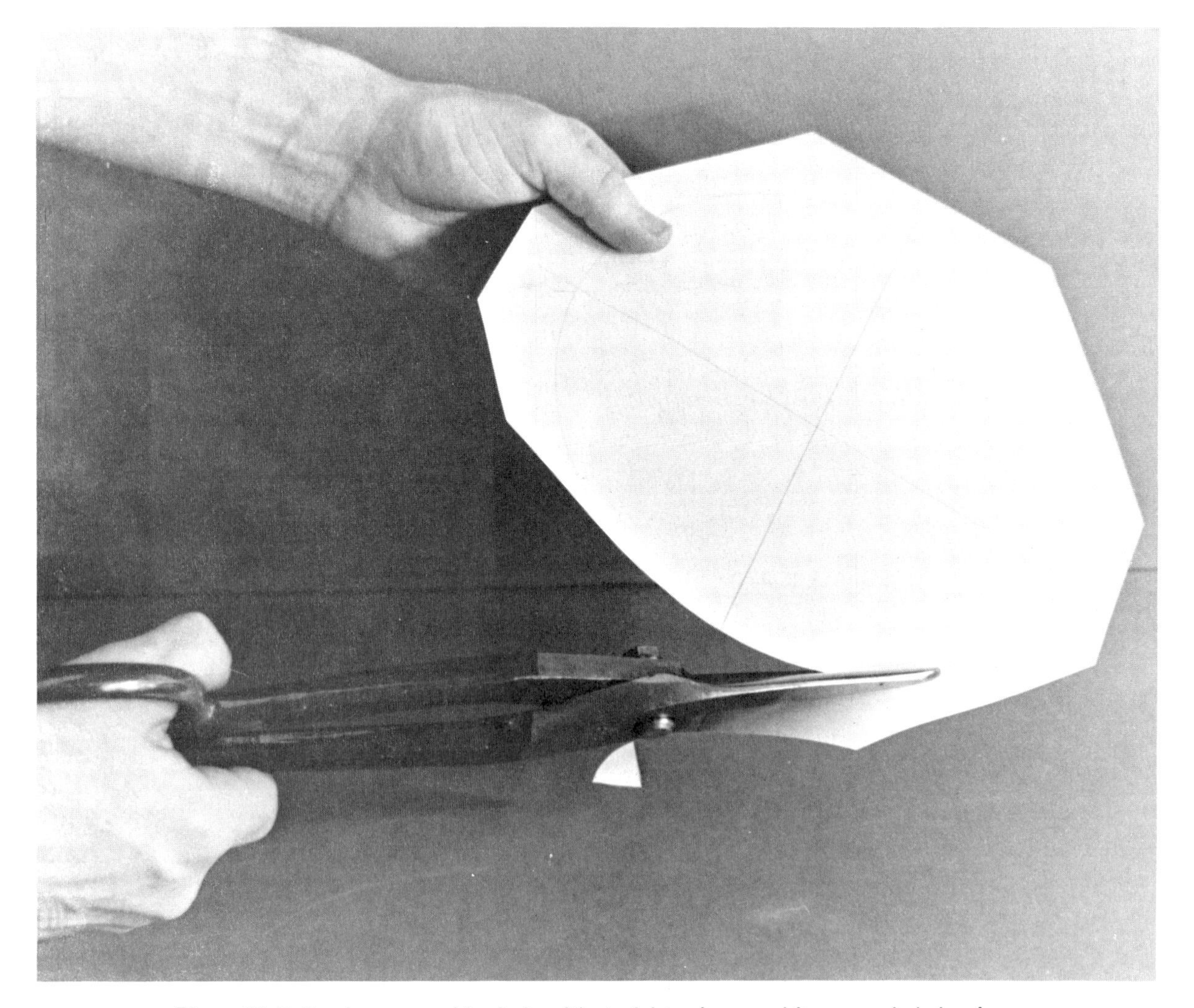

Figure 11-5 Cutting an outside circle with straight snips on white enameled aluminum.

2. Hold the metal in the left hand and the snips in the right hand. Rest the snips and the sheet on the bench if necessary.

3. Make a continuous cut, turning the metal as the cut is being made, figure 11-5. If possible, remove the waste material in one piece.

File off any jagged edges or slivers after cutting.

C. *Notching*

1. Select the proper snips. For mild steel, 22 gage and lighter, use either the straight or combination snips. For 16- to 20-gage mild steel, use the bulldog snips.

2. Grasp the snips in the right hand and the metal in the left hand. Rest the snips and the metal on the bench if necessary.

3. Open the blades and place the job between the blades, figure 11-6. The slit should be toward the right hand; and the part to be cut out, toward the left hand.

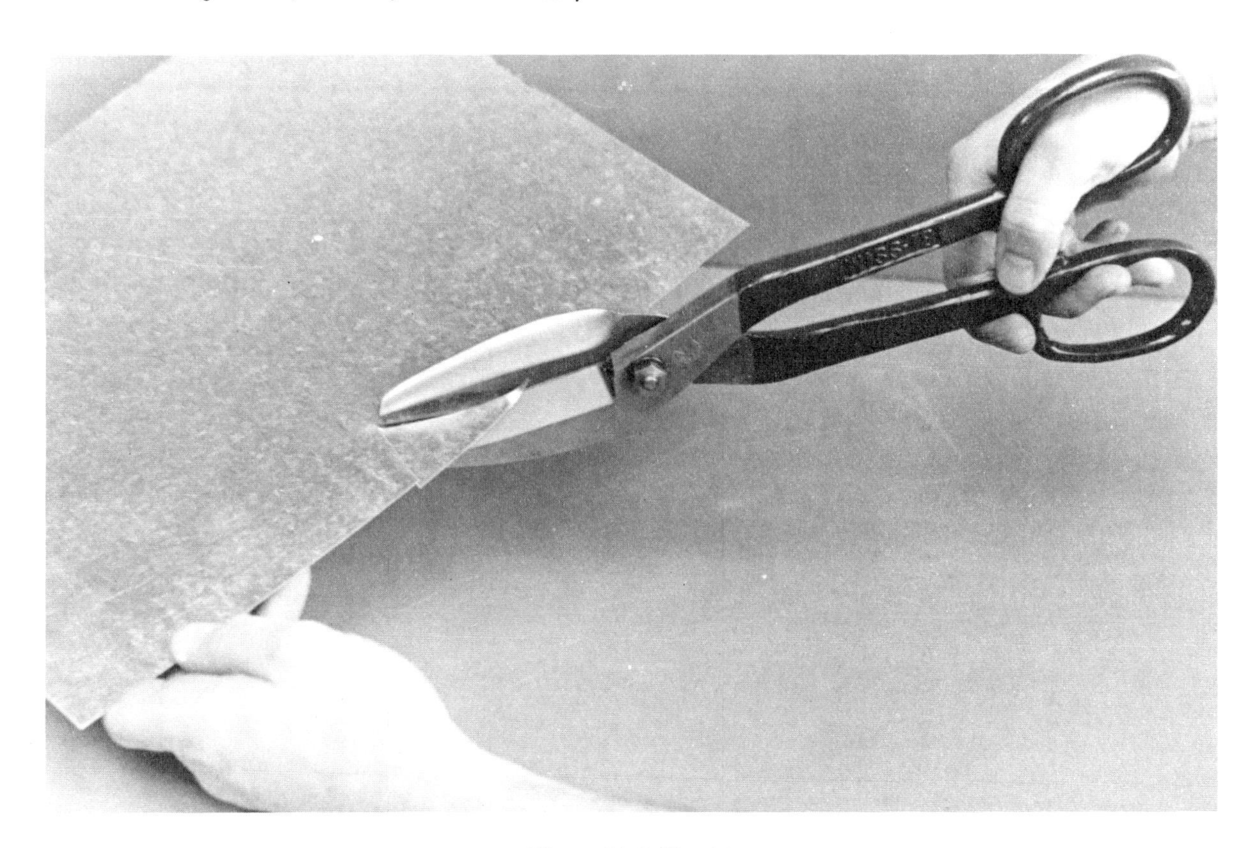

Figure 11-6 Notching.

The snips should be held at right angles to the sheet, and the blades must not extend beyond the notch line.

4. Make the slit by closing the blades.

5. Cut the opposite side of the notch.

Be careful that the piece being notched out does not fly toward fellow workers.

Double-Cutting Snips

Double-cutting snips, figure 11-7, as the name implies, makes two cuts at once to remove a ribbon of metal 1/8 inch wide from the sheet. They are so designed that the right- and left-hand pieces of metal being cut can lay flat. Thus, in cutting off sheet metal pipe lengths, neither side of the metal has to slide over the bottom blade as when using other snips.

Figure 11-7 Double-cutting snips. (Courtesy of Niagara Hand Tools.)

HOW TO USE DOUBLE-CUTTING SNIPS

1. Make a starting hole in the pipe on the layout line with a chisel or punch and hammer.
2. Open the cutting blade of the snips as far as possible and insert it in the starting hole.
3. Cut along the layout line, holding the snips in the right hand and rotating the pipe with the left hand so that the cut is always on the top surface, figure 11-8.

Figure 11-8 Using double-cutting snips.

4. Keep the snips at right angles to the pipe, and follow the layout line with the nose of the cutting blade.

File off any jagged edges or slivers after cutting.

CIRCULAR-CUTTING SNIPS

Circle Snips

Circle snips, figure 11-9, have curved blades and are used for making circular cuts. They come in the same sizes and capacities as straight snips and are available for right- and left-hand use.

Figure 11-9 Circle snips.

Hawk's Bill Snips

Inside and outside circles with small radii can be cut with hawk's bill snips, figure 11-10. The narrow, curved blades are beveled enough to permit sharp turns without buckling the material. Therefore, these snips are valuable for cutting holes in pipe, furnace hoods, and casings, and they are used in close quarters on cornice work.

Hawk's bill snips are available with cutting edges of 2 1/2 inches and an overall length of 11 1/2 inches and 13 inches. The capacity of the snips is 20-gage mild steel.

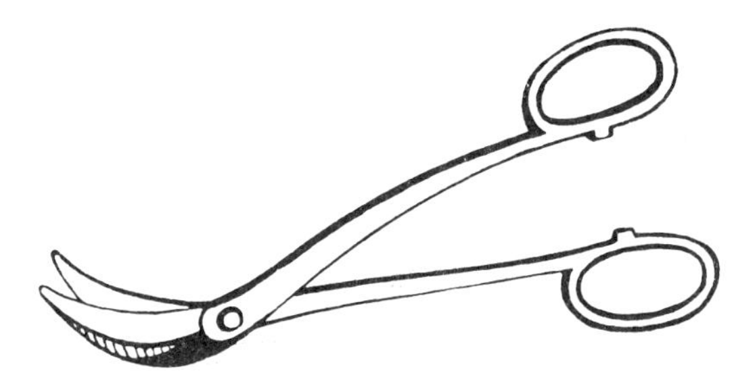

Figure 11-10 Hawk's bill snips.

Aviation Snips

Aviation snips have compound levers to lessen effort in cutting stock. These snips are used for cutting circular, square, and irregular patterns. The cutting blades are hardened so that hard material can be cut.

Aviation snips are made in straight, left- and right-hand types, figure 11-11. The upper blade of the right-hand snips is on the right, and the direction of cutting is to the left. Left-hand snips cut to the right. Both snips may be used with the right hand. Aviation snips are 10 inches long with a 2-inch length of cut. The capacity of the snips is 16-gage mild steel.

Double cutting and bulldog aviation snips are available. The bulldog style is used for notching and nibbling sheet metal up to 16-gage and for cutting stainless steel.

Newly designed aviation snips have molybdenum steel jaws and both cutting edges are serrated to avoid slippage in sharp radius work. The handles are color coded in keeping with industry standards: green cuts right, red cuts left, and yellow cuts straight.

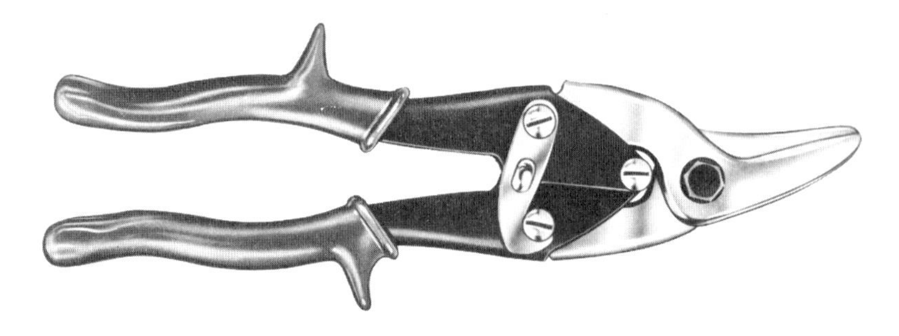

(A) Green handles cut to the right. (Courtesy of Stanley Tools.)

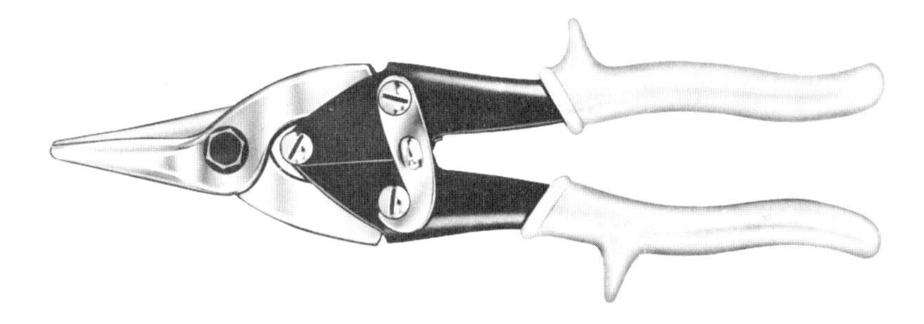

(B) Yellow handles cut straight. (Courtesy of Stanley Tools.)

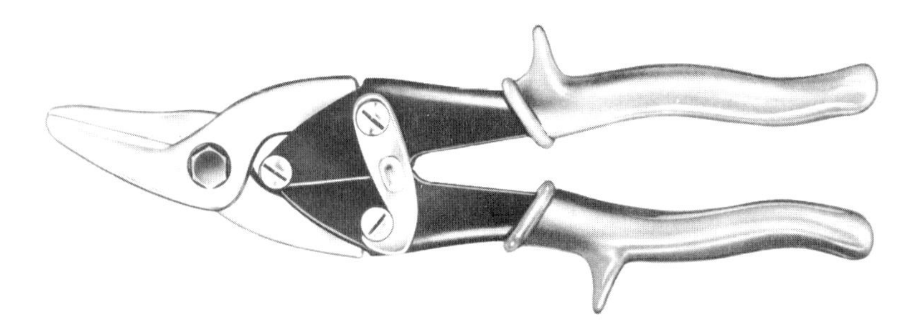

(C) Red handles cut to the left. (Courtesy of Stanley Tools.)

Figure 11-11 Aviation snips. (Courtesy of Stanley Tools.)

Trojan Snips

Trojan snips, figure 11-12, have slender blades and can be used for both straight and curved cutting. The blades are small enough to permit sharp turns without buckling the material. These snips are used to cut outside curves and may also be used instead of the circle snips, hawk's bill snips, or aviation snips when cutting inside curves.

The blades are made of forged, high grade steel. These snips are supplied in two sizes: one has a 2 1/2-inch cutting length and a 12-inch overall length, and the other has a 3-inch cutting length and a 13-inch overall length. Trojan snips have a 20-gage mild steel capacity.

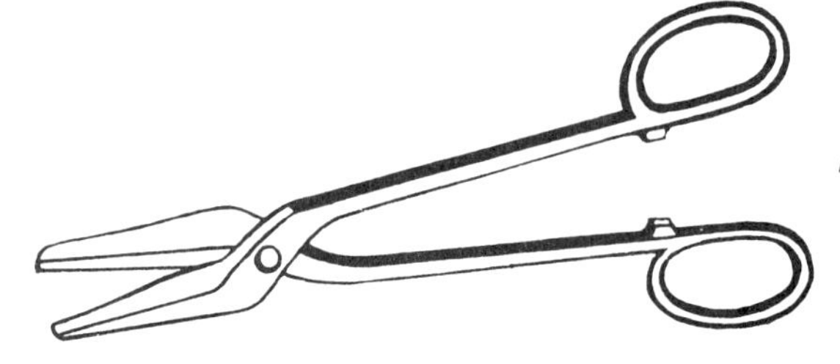

Figure 11-12 Trojan snips.

HOW TO MAKE AN INSIDE CURVED CUT WITH CIRCULAR-CUTTING SNIPS

1. Using a hammer and chisel, make two slits at right angles near the center of the hole to be cut, or punch a starting hole with a hollow punch after placing the metal on a lead cake or hardwood block.
2. Select the proper snips. Hawk's bill, aviation, or trojan snips are more convenient to make inside curved cuts than circular snips.
3. Insert the snips in the starting slit or hole from the top side of the metal.
4. Start cutting at the slit, gradually increasing the curve of the cut until the required opening is obtained, figure 11-13. When a large hole is to be cut, the center part can be cut out within 1/2 inch of the line with a chisel and hammer and then finished with the snips. This practice conserves material. If the left-hand aviation snips are used, the cutting is done to the right instead of to the left.

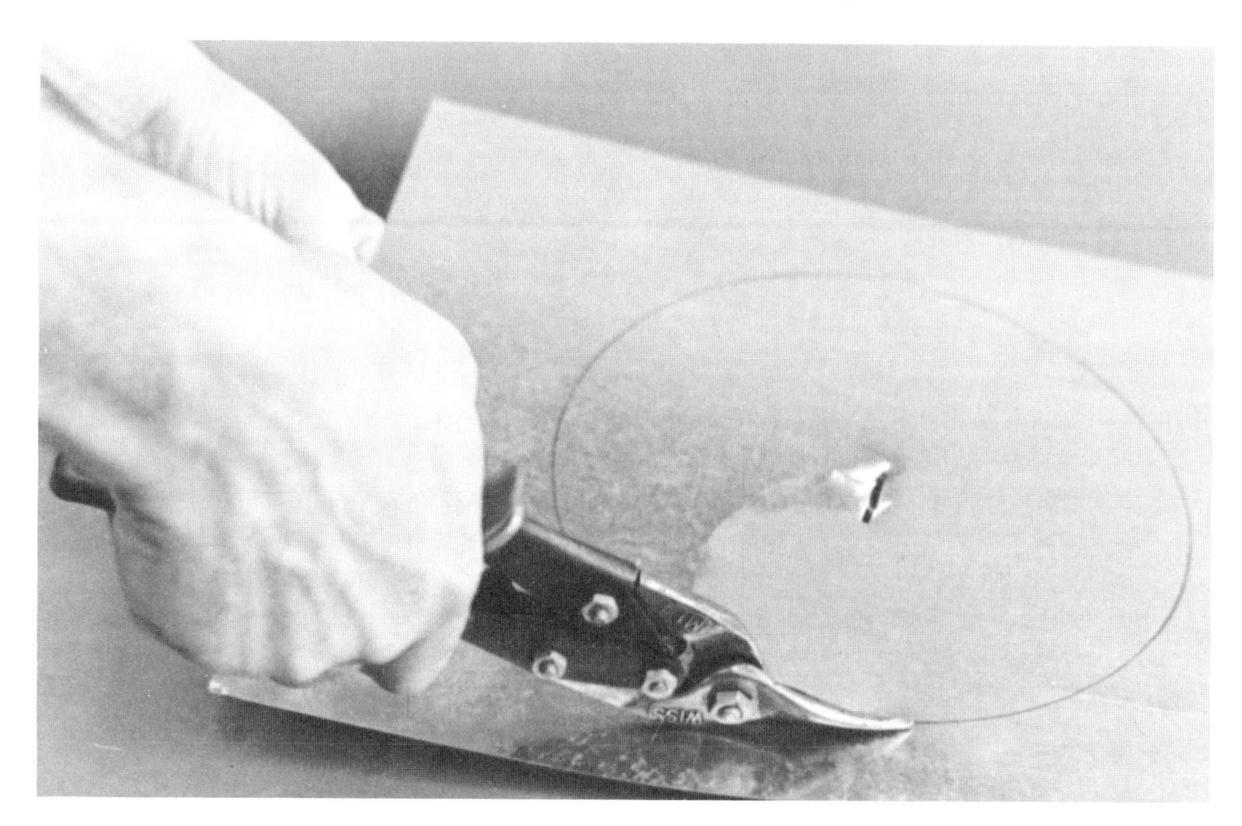

Figure 11-13 Cutting an inside circle with right-hand aviation snips.

File off any jagged edges or slivers after cutting.

GENERAL CARE

Hand snips should be properly oiled and adjusted to permit ease in cutting and to produce a surface which is free of burrs. If the blades bind or are too far apart, the snips must be adjusted.

HOW TO OIL AND ADJUST SNIPS

1. Oil the entire length of the blade and work the machine oil into the pivot bolt.

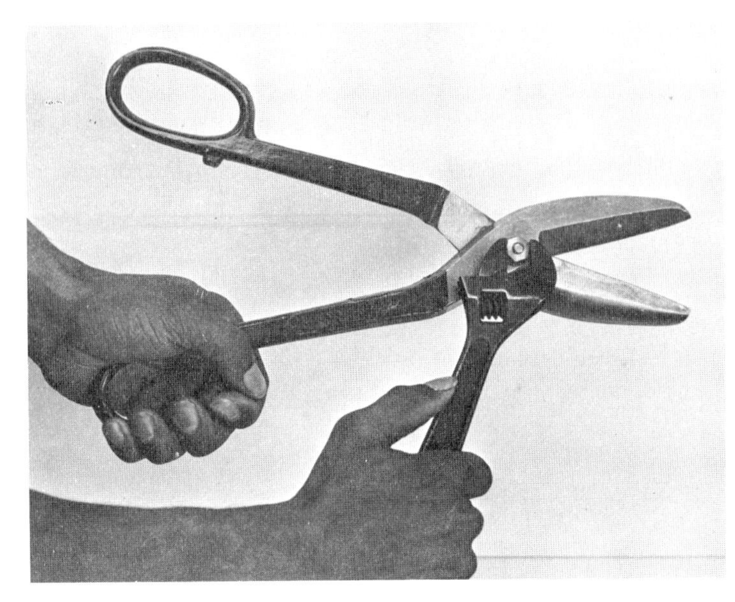

Figure 11-14 Adjusting the snips.

2. Open the snips, and tighten or loosen the nut with a small wrench until the correct clearance has been obtained, figure 11-14.

The blades must not be so tight that they bind excessively.

HOW TO SHARPEN SNIPS

1. Separate the blades by removing the pivot nut and bolt.
2. If the blade is chipped or dented, grind it to a smooth finish on a fine grinding wheel. Move the blade across the face of the wheel while grinding in order to cover the entire length of the blade. Hold the blade at an angle to match the original cutting edge.
3. For finish or light sharpening, stone the blade with a fine or medium bench stone. Place the sharpening stone on the bench and apply a light lubricant such as oil, benzene, or turpentine. Rub the cutting edge of the blade in a circular motion on the stone surface.

Be careful to keep the cutting edge parallel to the sharpening surface of the stone.

SUMMARY REVIEW

A. Place the answers to the following questions in the column to the right.

1. List seven different types of hand snips used in the sheet metal trade. 1. ______________________

____

____

____

____

____

____

2. State three characteristics of good hand snips. 2. ______________________
____
____

3. What are the two general classification of hand snips? 3. ______________________
____

4. What types of hand snips should be used to cut out the pattern for a pan 8 x 12 inches x 2 inches if 26-gage galvanized iron is required? 4. ______________________

5. A dovetail seam on a pipe pattern is to be cut out of 18-gage black iron. What type of snips should be used? 5. ______________________

6. A sheet of 18-gage black iron is to be cut into pieces 4 x 6 inches for practice welding. List three types of snips that can be used. 6. ______________________
____
____

7. On a heating job, 6-inch diameter round takeoffs must be cut into the 26-gage trunk line. List the tools to be used to make these cuts. 7. ______________________
____
____

8. The round takeoff pipe from the trunk line in (7) has to be cut to fit the crossover length. What snips should be used? 8. ______________________

9. In making a notch in a piece of stock what part of the cutting blade is used? 9. ______________________

10. In what direction do right-hand aviation snips cut? 10. ______________________

11. In which hand are left-hand aviation snips held? 11. ______________________

12. In order to make a cut with straight snips, what two precautions should be observed for each closing of the blades? 12. ______________________
____

B. Insert the correct word in each of the following sentences.

1. Snips are made of __________ steel.
2. Snips for cutting heavy gages have ________ handles or ___________ arrangements.
3. The maximum cutting capacity of large straight snips is _______ gage mild steel.
4. The bulldog snips can cut _______ gage mild steel.
5. The jaws of straight snips have a _____ straight from the cutting edge.

C. Underline the correct word in each of the following sentences.

1. Snips can be used to cut (wire, rod, sheet metal).
2. Double-cutting snips are especially useful in cutting (wire, sheet, pipe).
3. Aviation snips are used for (circles, 12-gage mild steel, heavy rod).
4. The blades of snips should be (adjusted, freed, sharpened) to produce a burr-free cut.
5. The blade of the snips should be held (perpendicular, parallel, at the original cutting angle) to the grinding wheel when sharpening.

UNIT 12 HAND HACKSAWS

OBJECTIVES

After studying this unit, the student should be able to

- Identify and describe the types of hacksaw blades and frames commonly used.
- Select the proper blade for the job.
- Use the hacksaw for cutting various metals properly.

The hand hacksaw is used for cutting materials by hand if power hacksaws and other cutting methods are not available or not convenient.

The sheet metal worker also uses a *keyhole* saw and a *hole* saw in cutting various materials on the job. The keyhole saw has an adjustable handle and three different blades for wood, metal, and plastics. The hole saw is used in the drill for cutting in wood or steel.

The hand hacksaw is one of the most abused tools in the shop as shown by excessive breakage of the saw blades. To prevent breakage and to make satisfactory cuts, it is necessary that the correct blade be selected for the job and that the hacksaw be used skillfully.

HACKSAW FRAMES

The two types of hacksaw frames are the pistol grip, figure 12-1, and the straight handle, figure 12-2. The pistol grip is the more practical, while the straight handle is used infrequently. Either type may be purchased with an adjustable frame to accommodate 8-, 10-, and 12-inch blades, or with a solid steel frame to hold a blade of a definite length. Sliding adjustable studs provide a means of setting the blade to cut parallel to or at right angles to the frame for long or deep cuts.

Tension is applied to the blade by turning the threaded handle in the straight handle frame; and in the pistol grip, by turning a wingnut.

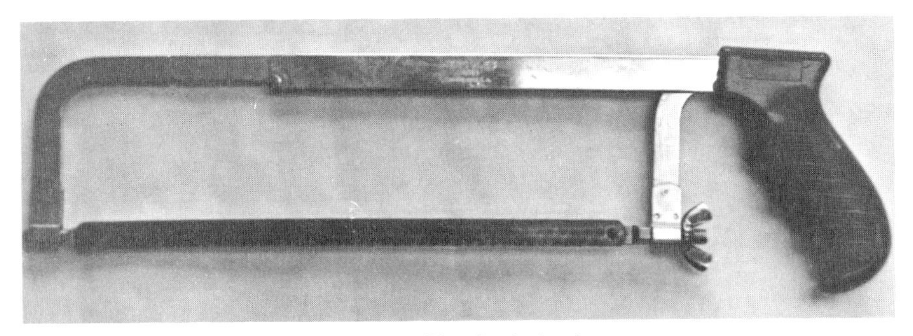

Figure 12-1 Pistol grip hacksaw.

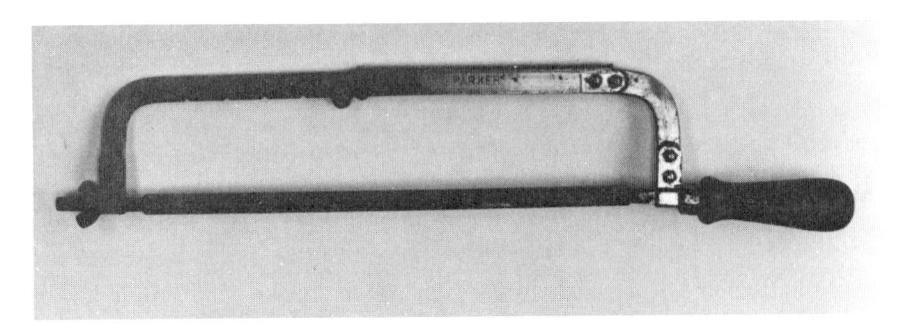

Figure 12-2 Straight handle hacksaw.

HACKSAW BLADES

Hacksaw blades are made of .90-percent to 1.10-percent carbon tool steel, tungsten alloy steels, and high speed steel. Steels containing .90-percent carbon harden to give maximum cutting speed without being too brittle. Adding 1.25-percent tungsten overcomes some of the brittleness.

Blades are hardened and classified as *all hard* and *flexible.* Only the teeth of flexible blades are hardened. This renders them less likely to break during cutting than the all-hard blades. Flexible blades are used to cut thin metal.

The length of saw blades varies from 8 to 12 inches. This length is determined by the distance between the centers of the holes in the blade. The thickness of hand hacksaw blades is approximately .025 inch. The width of the blade ranges from 7/16 to 9/16 inch.

The number of teeth per inch is the most important factor to be considered when selecting a blade for the work, figure 12-3. Blades are made with 14 to 32 teeth per inch. A special type of blade is available in which each tooth varies in size; the finer teeth are located at the front end of the blade. The graduation in teeth, often called *points,* facilitates making the starting cut.

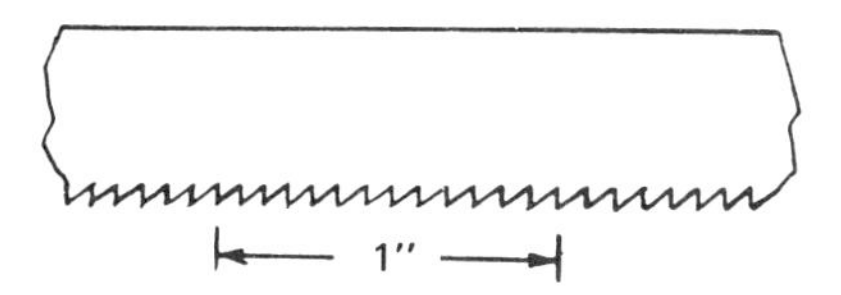

Figure 12-3 Teeth per inch.

To provide clearance for the saw blade and to prevent it from binding, the teeth are set to cut wider than the blade. The teeth can be arranged in an *alternate set,* a *raker set,* or a *wave set,* figure 12-4. On the alternate set blade, one tooth is bent slightly to the right and the next to the left. On the raker set blade, every third tooth is straight to clear the chips from the cut. On the wave set blade, short sections of teeth are bent in opposite directions. Wave sets are used on fine tooth blades.

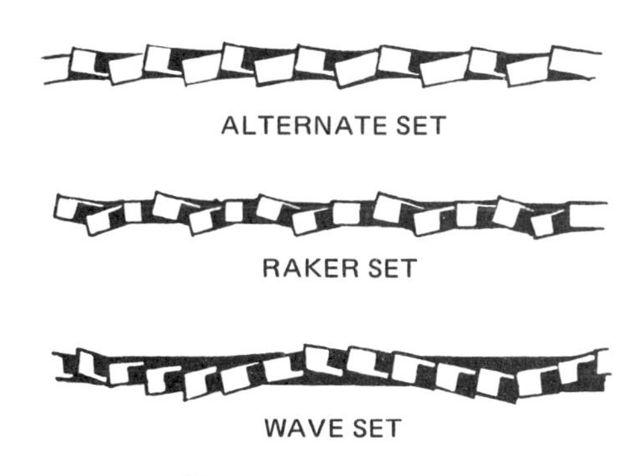

Figure 12-4 Set of blades.

CUTTING SPEEDS AND PRESSURE

The ease with which a piece of metal may be cut depends on the speed and the pressure applied to the saw. Manufacturers recommend a cutting speed of 40 to 50 strokes a minute. This permits the worker to saw without tiring and also gives him a chance to relieve the pressure on the return stroke. If the work is sawed too rapidly, the heat generated draws the temper of the blade, making it soft and useless. Enough pressure should be applied when cutting to prevent the blade from slipping. Slipping causes the cutting teeth to become glazed, thereby ruining the saw.

SELECTION OF HACKSAW BLADES

Table 12-1 shows the proper number of teeth per inch to use for the different materials and shapes of work. The result of using the wrong number of teeth is also shown. As a general rule, when selecting a blade for tubing or thin material, choose one which will have two or more teeth in contact with the surface being sawed.

TEETH AND USE	CORRECT	INCORRECT
14 per inch For large sections and mild materials	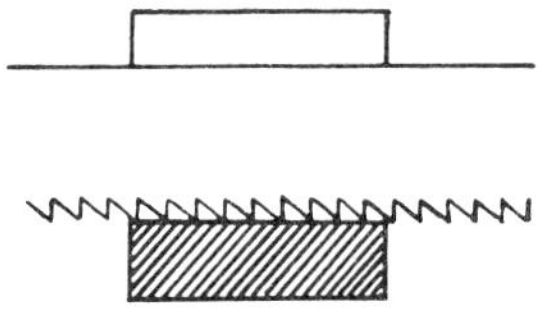 Good chip clearance	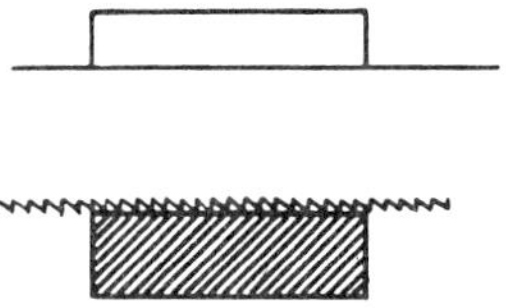 Teeth too fine No chip clearance RESULT: clogged teeth
18 per inch For tool steel, stainless steel, high carbon, and high speed steels.	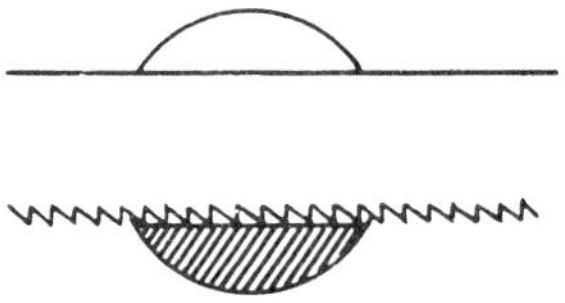	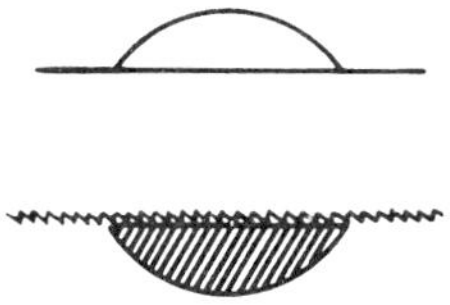
24 per inch For angle iron, brass, copper, iron pipe, BX, and electrical conduit	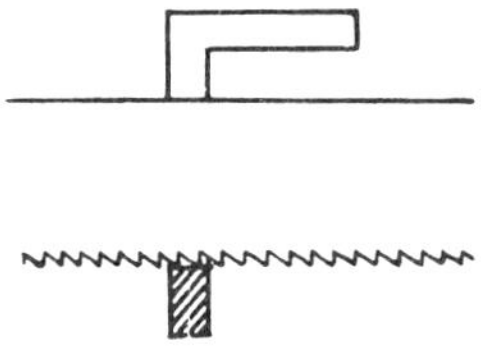 At least two teeth on a section	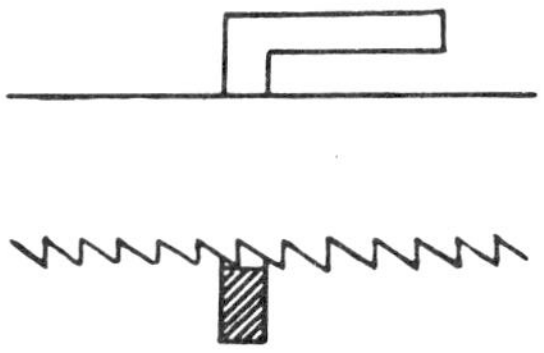 Teeth too coarse One tooth on section RESULT: stripped teeth
32 per inch For thin tubing and sheet metals	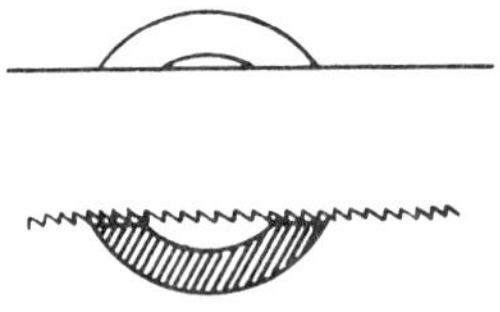	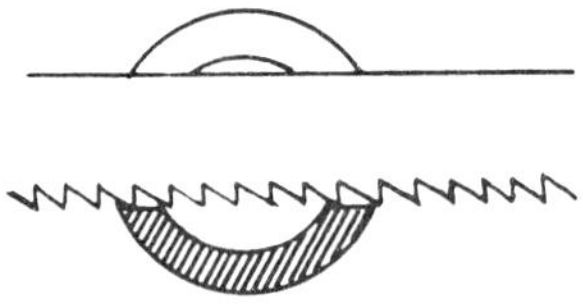

TABLE 12-1 SELECTION AND USE OF HACKSAW BLADES

HOW TO USE THE HAND HACKSAW

1. Select a saw blade for the job.
2. Place the blade in the frame so that the teeth point away from the handle and the saw cuts only on the forward stroke, figure 12-5.

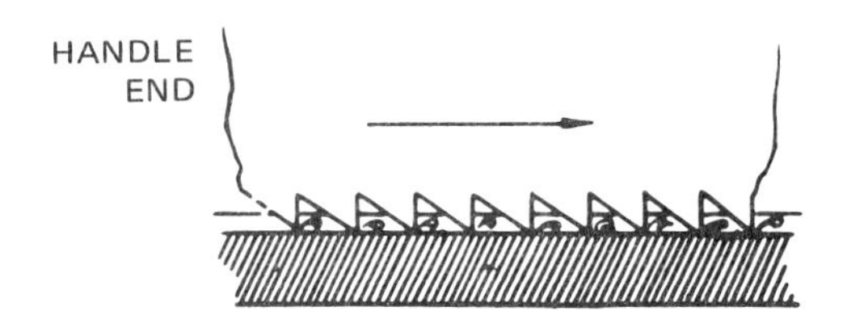

Figure 12-5 Direction of cutting.

3. Tighten the blade in the frame to prevent the saw from buckling and cutting off the mark.

Use judgment when tightening the frame to prevent breaking the blade, shearing the pins, or bending the frame. If too tight, the blade will *cant* (bend upward). If too loose, the blade will bend while being used and may break.

4. Clamp the work in the vise so that as many teeth of the saw blade as possible come in contact with the surface of the material to be cut, and in such a way that the saw cut line can be seen, figure 12-6. If the work has finished surfaces, use soft jaws in the vise to prevent marring. Place the work so that the cut line is approximately 1/4-inch from the vise jaws. This prevents the work from vibrating, figure 12-7. In addition, the saw cut line should be at right angles to the face of the vise jaws.

When band iron or sheet metal is to be cut, be sure that at least two teeth are in contact with the surface being cut. For instance, if a 32-teeth per inch blade is being used, metal 1/16-inch thick (2 x 1/32) is the thinnest metal that can be cut across the thin edge.

Figure 12-6 Proper position with maximum teeth in contact with surface.

Figure 12-7 Proper position of cut line to prevent the work from vibrating.

5. Indicate the starting point by nicking the surface with a file to break any sharp corner which might tend to strip the teeth. This mark also helps the beginner start the saw at the proper place.
6. Grasp the handle of the frame firmly with the right hand, figure 12-8; The front end of the frame is held by the left hand to help guide the saw when cutting.

Stand at the vise as shown in figure 12-8 with the left foot pointing toward the bench and the right foot close enough to the left foot to give the necessary balance.

Figure 12-8 Cutting with a hand hacksaw.

7. To start the cut, place the front end of the blade on the mark in the position shown in figure 12-9. Apply a little pressure and make the stroke by pushing the saw straight across the surface of the work. Release the pressure and return the saw to the starting position.

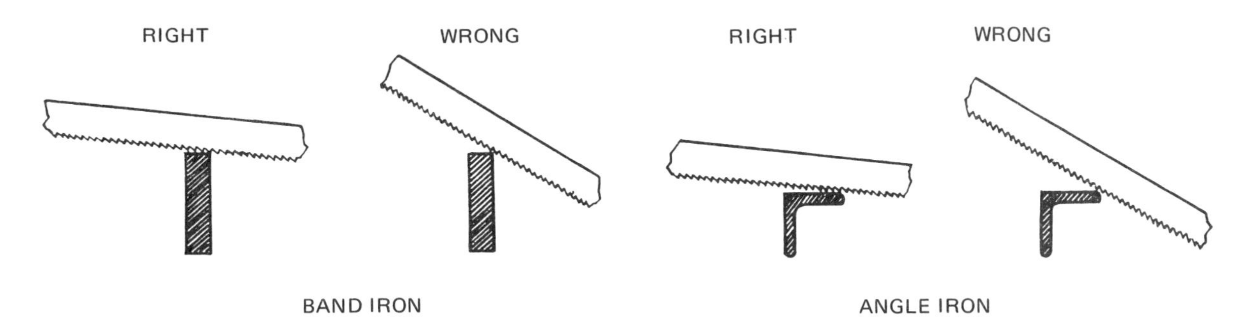

Figure 12-9 Starting the cut.

8. Repeat the process by adding a little pressure on the succeeding strokes. Continue the process at the rate of 40 to 50 strokes a minute for average work. No lubricant is necessary when cutting by hand. However, a little oil applied to the sides of the blade helps in making a deep cut.

The following points should be carefully noted while cutting to prolong the life of the saw blades:

a. Too much pressure should not be applied on the cutting stroke.
b. Make full length strokes to prolong the life of the blade and make the cutting easier.
c. Cutting too rapidly ruins the blade.
d. Make an even stroke so that the blade will not be twisted while sawing.
e. If the saw cuts off the mark, it should not be straightened by twisting. Start a new cut.
f. When a saw is broken in an unfinished cut, the cut should be resumed in another place on the work because the set of a new saw is thicker than that of a used saw. A new blade will break if forced into the old cut.

9. When the work is nearly cut through, use less pressure to prevent the saw teeth from catching.

10. Remove any burrs left by the saw with a smooth cut file.

11. Return the hacksaw to the place provided for it.

SUMMARY REVIEW

A. Place the answers to the following questions in the column to the right.

1. List the types of material-cutting saws used by the sheet metal worker. 1. ______________________

____

____

2. Identify the two types of frames commonly used on the hand hacksaw. 2. ______________________

____

3. List the common size hacksaw blades used in the field. 3. ______________________

____

____

4. How are hacksaw blades classified? 4. ______________________

____

5. What is the advantage of a flexible blade? 5. ______________________

6. Why are the teeth on a blade set to cut wider than the blade itself? 6. ______________________

____

7. List the various arrangements of teeth on hacksaw blades. 7. ______________________

____

____

8. What will happen if the worker saws a piece of steel with a hacksaw as fast as he can? 8. ______________________

9. What is the recommended speed in cutting with a hacksaw blade? 9. ______________________

10. What type of blade is least likely to break? 10. ______________________

11. Blades can be purchased with varying numbers of teeth per inch. List four such blades. 11. ______________________

____

____

____

12. What blade (teeth per inch) should be selected to cut a stainless steel rod? 12. ______________________

13. In making an iron frame with 1 x 1 x 1/4 inch steel angle, what blade should be selected to cut the miters? 13. ______________________

14. On which stroke should pressure be applied when cutting with a hacksaw? 14. ______________________

15. If, when cutting with a hacksaw, the worker notices that the blade bends, what is wrong? 15. ______________________

16. If, when making a cut, the blade runs off the line, what should the worker do? 16. ______________________

B. Insert the correct word in each of the following sentences.

1. The hacksaw blade can be adjusted to cut __________ to or __________ to the material.
2. Blades are hardened and classified as ______________ or ______________ .
3. Rapid cutting with the hacksaw generates __________ that draws the temper.

C. Underline the correct word in each of the following sentences.

1. To cut angle iron, a blade with (14, 18, 24, 32) teeth per inch should be selected.
2. The teeth on a hacksaw blade should always be placed in the frame so that the teeth point (towards, away from, at right angles to) the handle.
3. Always place band iron or sheet metal in a vise so that at least (5, 3, 4, 2) teeth on the blade are in contact with the surface.

UNIT 13 *COLD CHISELS*

OBJECTIVES

After studying this unit, the student should be able to

- Identify and describe the types of cold chisels.
- Select the proper cold chisel for the job.
- Use and sharpen a cold chisel properly.

The cold chisel is a wedge-shaped tool used to cut metal softer than itself. *Chiseling* is the process of removing or cutting metal by means of a cold chisel and a hammer or a pneumatic gun. The pneumatic gun is used when a good deal of metal is to be removed.

Cold chisels are usually forged from octagonal tool steel containing approximately .80-percent carbon. This steel is tough enough to withstand the blow of the hammer, and yet the cutting end can be hardened and tempered to prevent the tool from becoming dull rapidly.

TYPES OF CHISELS

A variety of chisels is available for different kinds of work. The common types are flat, cape, diamond point, and round nose, figure 13-1. Chisels are ordered by the width of the cut and the overall length. Sheet metal workers generally use the flat chisel, sometimes called simply the cold chisel.

A flat chisel is used for cutting sheet metal, band iron, and wire, and for removing stock from flat surfaces, a process known as *chipping.* It is also used to cut holes in sheet metal jobs, to make starting slits for snips, and to cut off the heads of rivets and bolts. The other chisels are used mainly for cutting narrow grooves of different shapes and for cutting corners.

Figure 13-1 Hand cold chisels. (Courtesy of Stanley Tools.)

CUTTING ANGLES AND EDGES

A sharp, correctly ground chisel should be used at all times. The cutting angle of the chisel is determined by the strength of the material being cut. Chisels used on hard metal require a strong cutting edge with an included angle of 70 degrees, figure 13-2. This angle can be decreased to 60 degrees for softer metals because less pressure is needed in the cutting process.

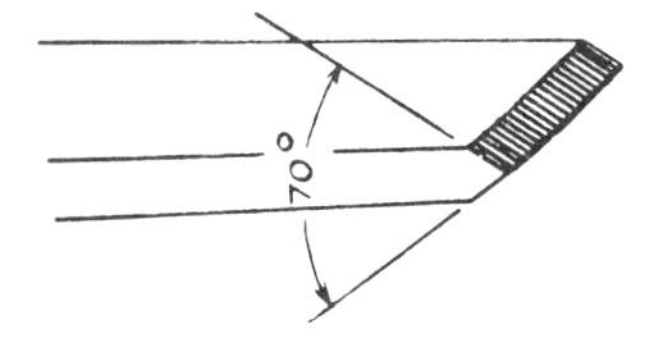

Figure 13-2 70-degree included angle.

Figure 13-3 Mushroomed head.

The head of a chisel should be dressed by grinding to its original shape (beveled slightly) if it becomes mushroomed while in use, figure 13-3. It is dangerous to use a chisel with a mushroomed head because the burrs may fly off and cause injury. Always return a chisel to the tool crib in good condition.

When the chisel does not maintain a sharp cutting edge, it should be rehardened and ground to the proper cutting angle. Cracks and breaks in chisels are caused by improper forging or heat treating. Cracks indicate that the tool was either forged at too cold a temperature or heated too much when hardening. Breaks are usually caused by poor tempering.

The flat chisel gives better results if ground with a slightly convex cutting edge, figure 13-4. With a convex cutting edge, there is less tendency for the corners to dig into the surfaces being chiseled. This method of grinding also centralizes the pressure exerted on the tool. When the head of the chisel is struck, the impact is taken up by the center of the tool where it has more material to withstand the strain.

Figure 13-4
Convex cutting edge.

HOW TO DRESS A GRINDING WHEEL

1. Support the dresser on the tool rest so that the point of contact between the wheel and the dresser is slightly above the center of the wheel, figure 13-5.
2. Pass the dresser back and forth across the face of the wheel while it is in motion.
3. Form the face of the wheel to an even surface which is approximately square with the sides of the wheel.
4. Stop the grinder and inspect the face of the wheel to see that all shiny spots are removed and that the pores of the wheel are clean.

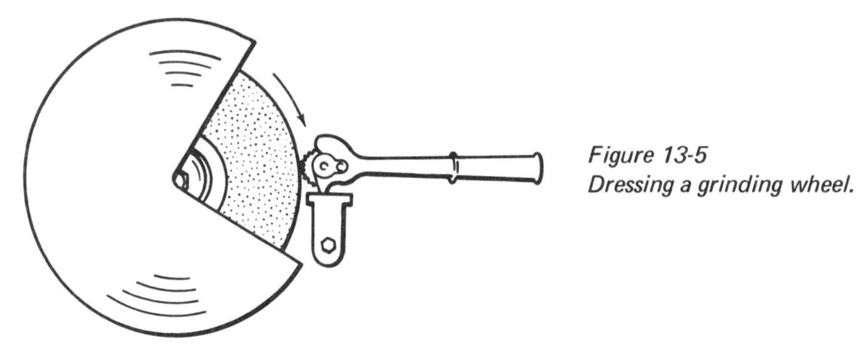

Figure 13-5
Dressing a grinding wheel.

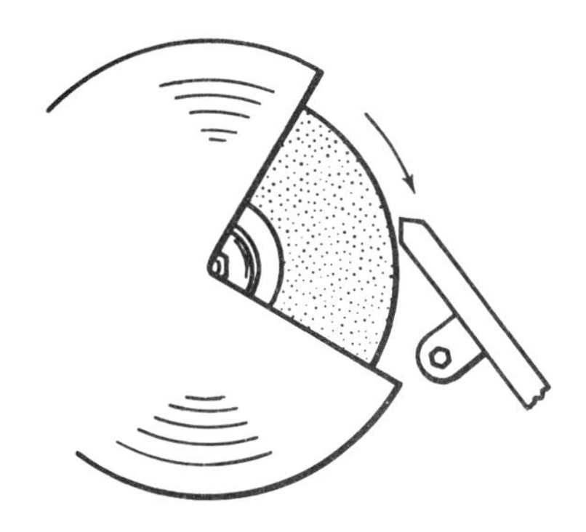

Figure 13-6 Grinding a flat chisel.

HOW TO GRIND A FLAT CHISEL

1. Adjust the tool rest so that an included angle of 70 degrees is formed.
2. Hold the flat part of the chisel on the rest and lightly press the beveled surface against the grinding wheel face, figure 13-6.
3. Rotate the chisel slightly while grinding to form a convex cutting edge on the tool.

Too much pressure should not be applied while grinding a chisel because the heat generated may draw the temper. The chisel should also be cooled periodically during grinding by immersing it in cold water.

4. Turn the chisel over to the other side and repeat steps 1 through 3.

HOW TO CUT LIGHT SHEET METAL

Select the proper chisel and a riveting or ball peen hammer. For ordinary work, a 3/4-inch flat chisel and a 1-pound hammer are used. Prepare the chisel as necessary. Obtain vise jaws or angle irons and protective goggles.

Make sure that the head of the chisel is not mushroomed because particles may break off and cause injury.

1. Place the cutting line on the stock flush with the edges of the angle irons and secure the work and angle irons firmly in the bench vise. The angles irons prevent the chisel from cutting into the vise.
2. Hold the stock of the chisel firmly enough to guide it as shown in figure 13-7.

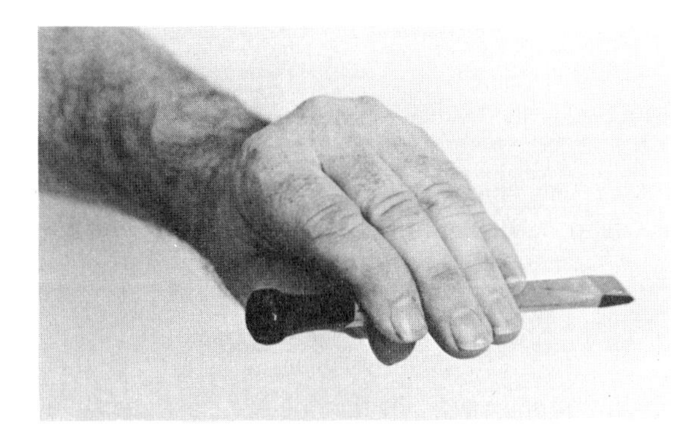

Figure 13-7 The proper way to hold a chisel.

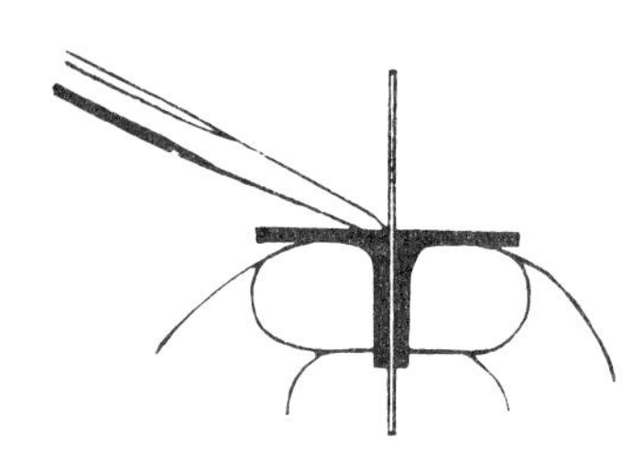

Figure 13-8 Proper placement of chisel cutting edge.

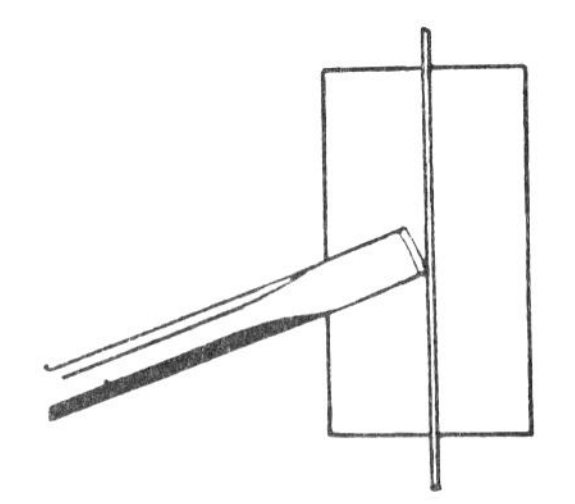

Figure 13-9 Chisel held at angle to work.

3. Starting at the center of the cut line, place the cutting edge of the chisel as shown in figure 13-8.

Hold the chisel at an angle to the job as in figure 13-9.

4. Grasp the hammer near the end of the handle so that it can be swung with an easy forearm movement, figure 13-10.

Figure 13-10 Cutting light metal in a vise.

5. Strike the head of the chisel with a sharp blow.

Watch the cutting edge of the chisel and not the head in order to direct the cut better. With practice, one can learn to strike the head of the chisel without looking at it.

6. Reset the cutting edge of the chisel on the cut line and repeat steps 4 and 5 until the cut is completed.

Begin chiseling at the center of the line to be cut and follow the line each way to prevent tearing the metal at the corners.

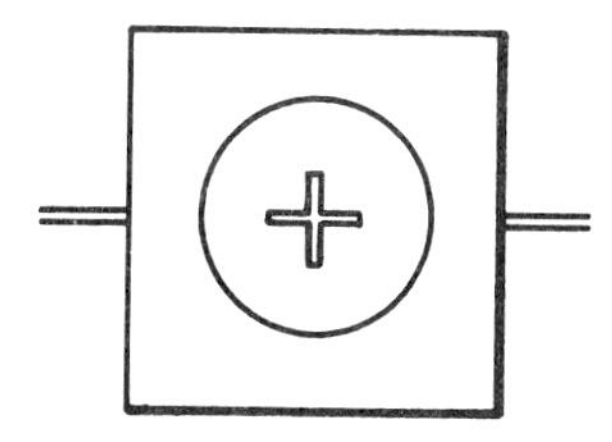

Figure 13-11 Starting slit.

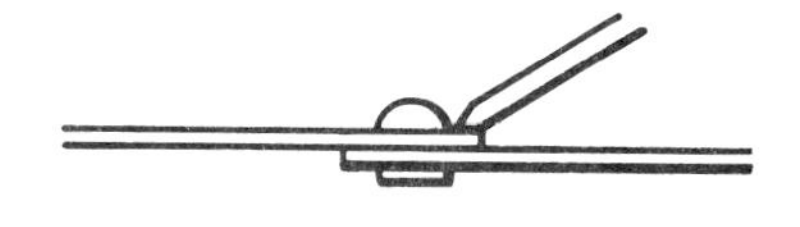

Figure 13-12 Removing rivet with a cold chisel.

HOW TO MAKE A STARTING SLIT

1. Place the job across an opening or at the end of the bench.
2. Cut a cross with the flat chisel and a hammer, figure 13-11.

HOW TO REMOVE A RIVET

1. Place the riveted job so that one end is held firmly against a solid object.
2. Hold the flat cold chisel at the rivet head with the cutting edge parallel to the stock surface, figure 13-12.
3. Strike the chisel with a hammer until the rivet head is removed.

SUMMARY REVIEW

A. Place the answers to the following questions in the column to the right

1. What is the purpose of a cold chisel? 1. ______________
2. List four commonly used cold chisels. 2. ______________
   ____
   ____
   ____
3. If a worker asks for a 1/2-inch cold chisel, what is he referring to? 3. ______________
4. List four uses for a flat cold chisel. 4. ______________
   ____
   ____
   ____
5. What three cold chisels may be used to cut a narrow groove in a piece of stock? 5. ______________
   ____
   ____
6. What is the proper cutting angle on a cold chisel for cutting hard metal? 6. ______________
7. What shape cutting edge is best for the flat chisel? 7. ______________
8. What happens to the cutting edge if it becomes hot enough to turn blue during grinding? 8. ______________

9. When is it dangerous to use a cold chisel without regrinding? 9. ______________

10. What is wrong with a cold chisel if it cracks or breaks along the cutting edge in use? 10. ______________

11. What type and size of cold chisel should be selected to cut a 2 x 4 inch opening in 16-gage black iron? 11. ______________

12. Where should the worker focus his eye when striking a chisel with a hammer? 12. ______________

13. Where should the cut be started on sheet stock? 13. ______________

14. A 4-inch square hole is to be cut in a 24-gage galvanized iron duct. How should the cut be started? 14. ______________

B. Insert the correct word in each of the following sentences.

1. Cold chisels are usually __________ from octagonal tool steel.
2. The processes of removing or cutting metal by means of a cold chisel and hammer is called__________
3. The purpose of a convex cutting edge on a cold chisel to ________ the pressure exerted.
4. The head of a chisel should be dressed by grinding if it becomes ___________while in use.

C. Underline the correct word in each of the following sentences

1. A cold chisel is (round, straight, curved, wedge shaped).
2. A cold chisel is used on construction jobs in which (accuracy, efficiency, rapidity) is not important.
3. The head of a chisel should always be (mushroomed, beveled slightly, straight).
4. In cutting with a cold chisel, the worker should always watch the (chisel head, hammer head, chisel cutting edge).

UNIT 14 *FILES*

OBJECTIVES

After studying this unit, the student should be able to

- Identify and describe the common classifications of files.
- Properly care for and use a file.
- Select the proper file for the job.

The file is a cutting tool which has a large number of teeth cut diagonally on the face. Most files are made of a high grade tool steel and are hardened and tempered. Files are manufactured in a variety of shapes and sizes adaptable to the job at hand. They are identified according to the characteristics of the cross section, the general shape, or the particular use. The cuts of files must be considered when selecting them for various types of work and materials.

Files are used by the sheet metal worker to square the ends of band iron and other shapes of iron, to file rounded corners on heavy iron, to remove burrs, slivers, and fins from sheets or jobs, to straighten uneven edges, to file holes and slots, to smooth soldered joints, and to file soldering coppers. Filing also smoothes rough surfaces.

It is difficult to file properly, and the worker must give careful attention to details and practice to become proficient. The proper kind of a file must be selected to do a workmanlike job. The file must be clean; and the handle, tight. The most suitable method of filing depends on the size, shape, and position of the particular job.

The cutting action of a file produces small chips called *filings.* These small particles of metal frequently become wedged between the teeth of the file due to the pressure exerted. When drawn across the work, the embedded filings scratch the surface and impair the free cutting action of the file. Therefore, frequent cleaning is necessary to produce a smooth surface free from scratches and to obtain the maximum efficiency of the file. A file should not be cleaned by rapping it on a hard surface because it may chip the teeth or break the file.

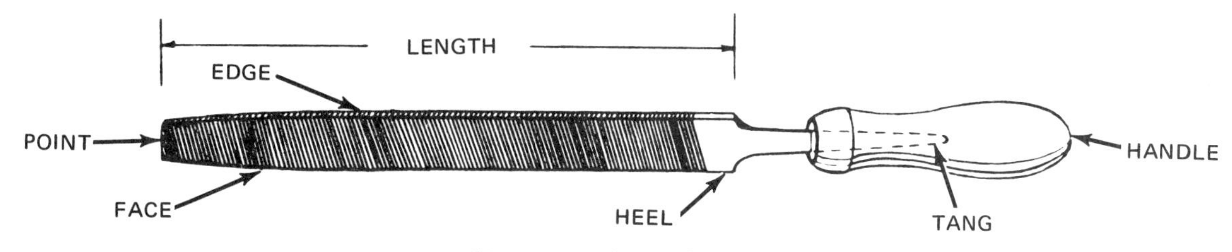

Figure 14-1 Parts of a file.

PARTS

The parts of a file are known as the tang, heel, face, edge, and point, figure 14-1. The length of a file is measured from the heel to the point, not including the tang. The parts of a file are defined as follows:

Tang: That part of a file to which the handle is fitted.

Heel: A small portion next to the tang which has no teeth. The kind of cut is usually stamped on this part.

Face: That surface of the file upon which the teeth are cut. It is also known as the *cutting surface.*

Edge: The narrow portion between the faces. When the edge is left smooth (without teeth), it is called a *safe edge file.*

Point: The point is the tip of the file.

Handle: Wood or composition formed for convenient use and fitted to the tang of the file proper. The size of the file and the nature of the work determine the size of the handle necessary for proper balance. Handles have a hole drilled in them to guide the tang and also prevent splitting of the handle.

The file should not be used with a loose handle or without a handle because the tang may be driven into the palm of the hand, causing injury.

SHAPES

The proper selection of a file for the job requires a knowledge of the various shapes and their applications, figure 14-2. The flat, mill, hand, and pillar files have about the same cross-section, but the shape of the sides differs. A flat file is used to file flat surfaces. The mill file is an all-purpose file especially adapted for finish filing. The hand and pillar files have safe edges to prevent cutting both sides of a corner at the same time. Half-round files are used on concave surfaces and large radii, while round files are better suited to smaller radii. The square, round, three-square, and half-round files can be used to file holes of various shapes.

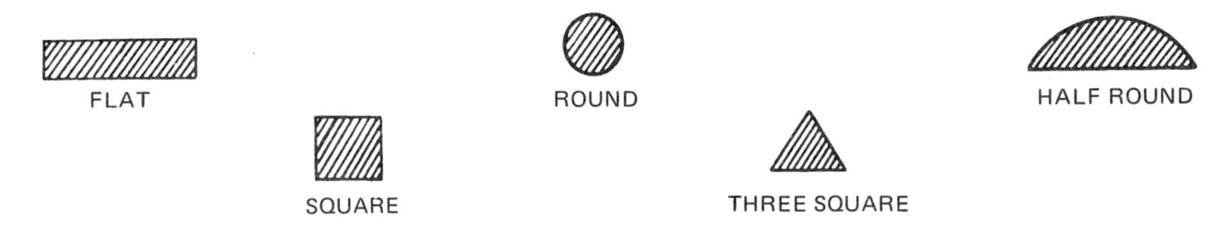

Figure 14-2 Shapes of commonly used files.

CUTS

Files are made in two types of cuts, the single cut and the double cut. The teeth of files vary in fineness (smoothness) and arrangement. Both single-cut and double-cut types have teeth cut parallel to each other for the length of the file.

Single-cut files have a single set of teeth cut at an angle of 65 to 85 degrees, figure 14-3A. These are used for filing thin work (sheets) and for finishing. All standard mill files are single cut.

Double-cut files have two sets of teeth that cross each other, figure 14-3B. One set is cut at an angle of 40 to 45 degrees; the other set is cut at an angle of 70 to 80 degrees. One set is cut deeper than the other. The teeth are a series of points. Because double-cut files remove material faster than single-cut files, they are used for roughing.

Figure 14-3 Single and double cut files.

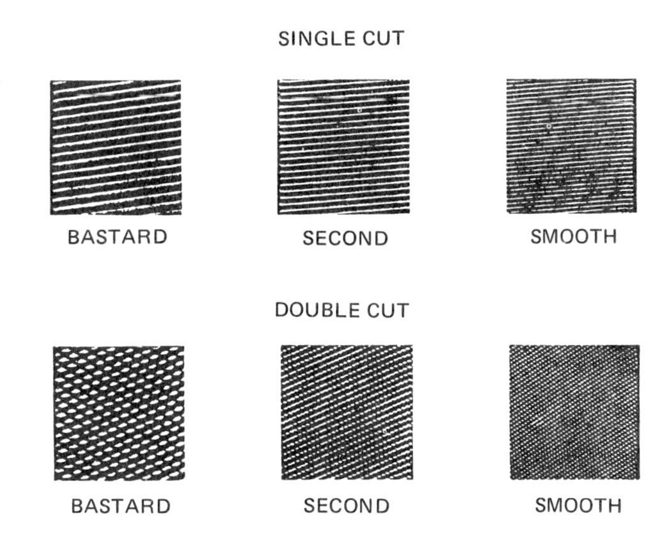

Figure 14-4 Cuts of files.

Figure 14-4 shows the various degrees of fineness and coarseness in the double-cut and single-cut types of files. The smoother files are used for finishing, and the rougher files for roughing and cutting soft materials.

A special file called the lead float file files soft metals such as lead and solder. This file has coarse, single-cut teeth, usually inclined at a sharp angle to permit chip clearance.

Vixen files have curved teeth and are used for filing lead, solder, aluminum, and copper. These files leave a smooth surface on the work. Because the teeth have ample clearance, they do not clog. Vixen files may be obtained in many of the same shapes as other files and fitted with the conventional wooden handles, or they may be obtained in special shapes and attached to a holder.

The Surform® file, figure 14-5, has a straight blade with holes that allow the shavings to pass through it and thus prevent clogging.

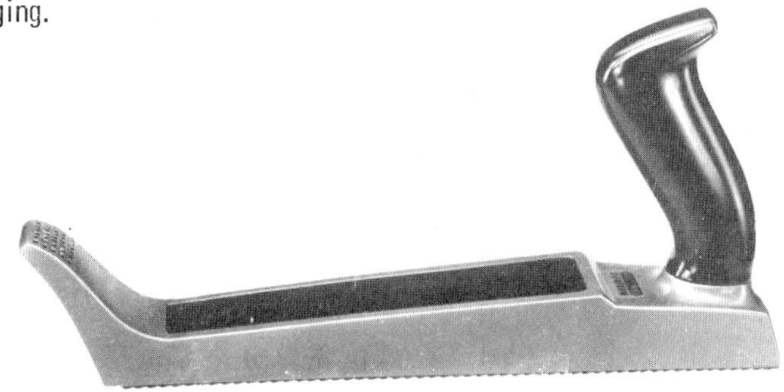

Figure 14-5 Surform® file. (Courtesy of Stanley Tools.)

HOW TO CARE FOR FILES

1. Always chalk a new file to keep the teeth from filling up. (See page 81.)
2. Break in a new file on a broad surface since the teeth are apt to catch and break if broken in on narrow edges.
3. Hold the work or sheet rigidly so that it does not chatter to avoid breaking the teeth of the file.
4. When filing, raise the tool slightly on the return stroke. The teeth are designed to cut on the forward stroke only.

5. Do not file rapidly because cutting is not efficient.
6. Clean files frequently since particles tend to lodge between the teeth, reducing the cutting ability of the file.
7. File light gage sheet lengthwise whenever possible.
8. Never use a file as a hammer or a pry bar. Files chip upon impact and break very easily. Flying chips are dangerous.
9. Never heat a file since it removes the temper.
10. Store files in a special place. Contact with other tools damage the teeth, thus reducing efficiency. A rack which keeps the files separated protects the tools.
11. Use the lead float file for filing solder and soldering coppers.
12. Fit files with a handle that has a metal ferrule whenever possible. This prevents splitting and increases safety.
13. Tap handles onto the tang until a snug fit is made. Nothing is gained by driving a handle on all the way.

HOW TO CLEAN A FILE

1. Use the wire brush side of the file card and brush parallel with the file teeth to remove particles of metal, figure 14-6.
2. To remove remaining particles of metal which are pinned between the teeth, push the filings out with the wire scorer on the other side of the card.

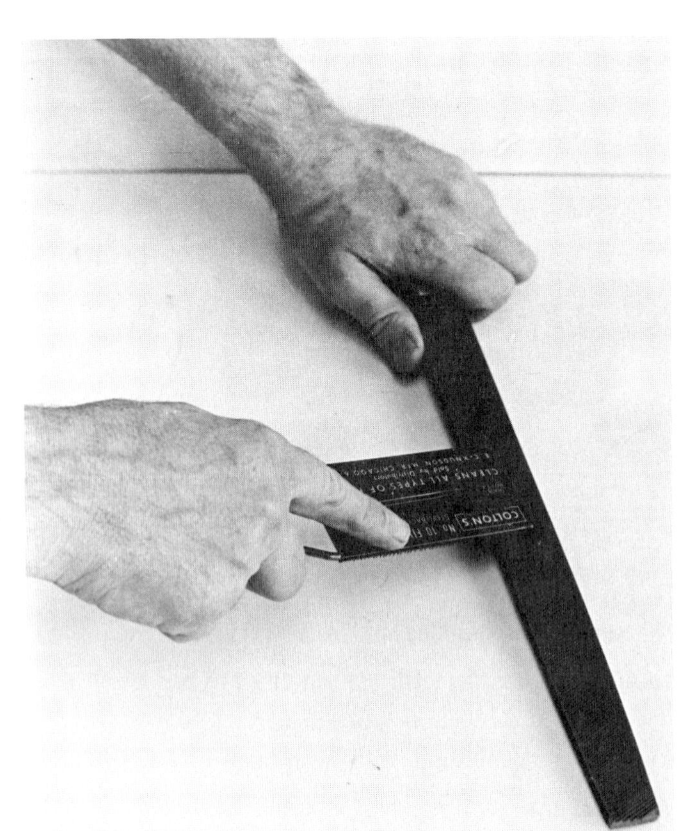

Figure 14-6
Cleaning a file with a brush.

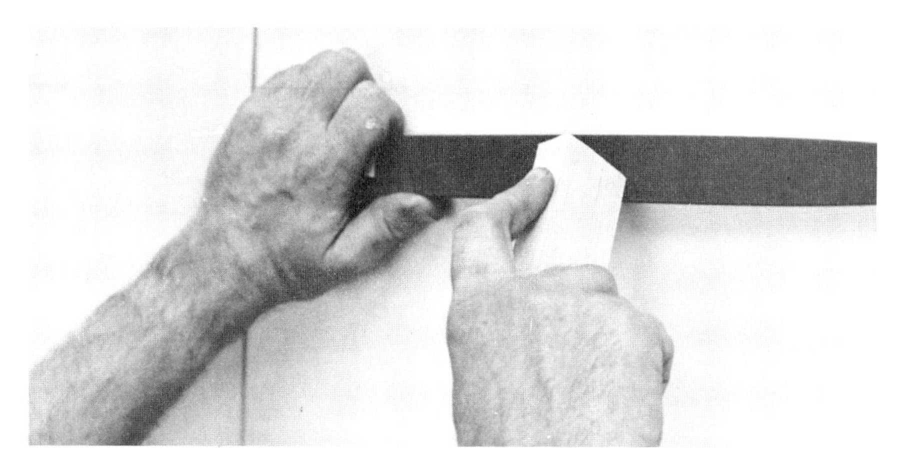

Figure 14-7 Cleaning a file with a wood block.

3. To remove soft metal particles from the file, use a piece of hard wood. Push the block across the file parallel with the teeth so that an impression of the file teeth is cut into the block. This action dislodges the particles from the teeth, figure 14-7.
4. If a smooth finishing cut is desired, rub chalk lengthwise on the file. This practice keeps chips from clogging the teeth and scratching the work because the chalk fills the spaces between the teeth to prevent particles from clinging to the file.

HOW TO FILE

1. Select the proper file.

A file should always be fitted with a handle to prevent the point from being driven into the hand. If the handle is loose, tighten it by tapping the wooden handle with a mallet.

2. Clamp the work to be filed in a vise. If the work has finished surfaces, use soft jaws in the vise to prevent marring the surfaces. When possible, the work should be held at the level of the elbow of the workman as he files. This position enables him to get the full swing of his arms from his shoulder.
3. Grasp the file in the right hand with the thumb resting on top of the handle as shown in figure 14-8. The thumb is placed on top of the handle to assist in guiding the file.

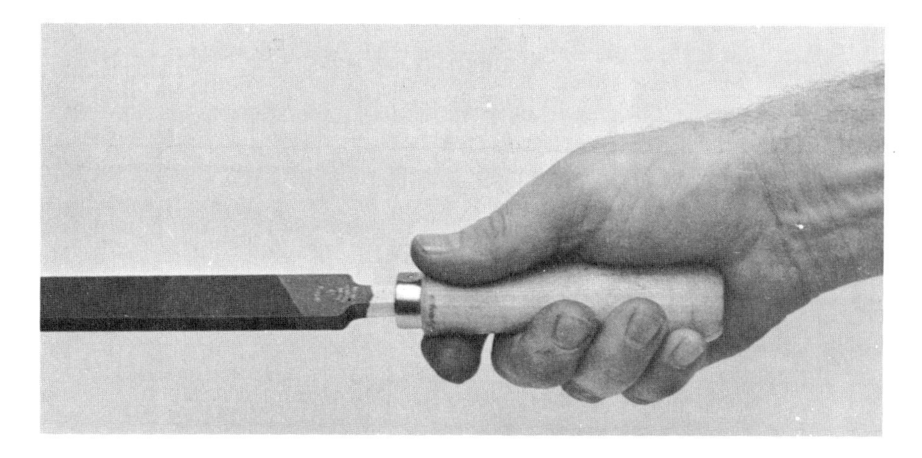

Figure 14-8 The correct way to grasp a file.

Figure 14-9
Correct position while filing.

4. Stand at the vise and hold the file parallel to the surface to be filed, figure 14-9. The point of the file should be grasped with the thumb and the first two fingers.

5. Push the file and bear down on the forward stroke.

CAUTION Do not rock the file to prevent an uneven surface.

6. Release the pressure and return the file to the original position for the next stroke.

If the pressure is not released, the teeth will wear excessively or be damaged.

7. Test the work frequently with a square or scale to determine whether the filing is straight or square, figures 14-10 and 14-11.

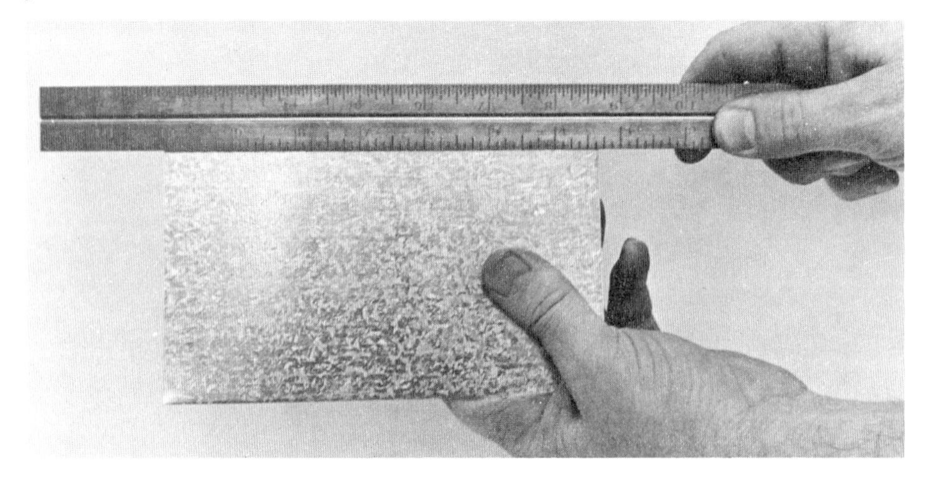

Figure 14-10 Testing for straightness with blade of combination square.

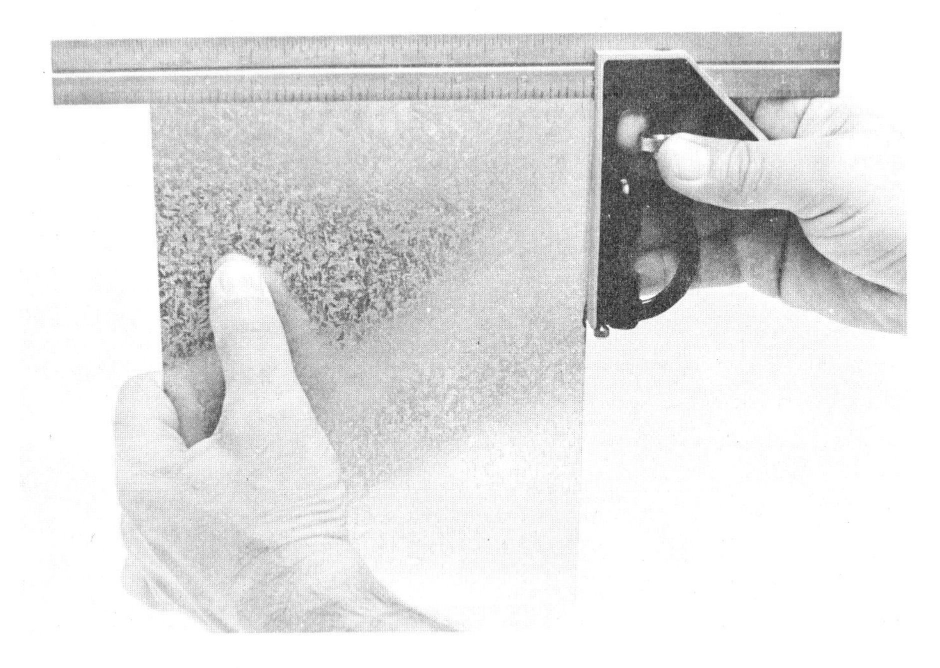

Figure 14-11 Testing for squareness with the square head.

8. When finish filing, use a fine-cut file. Grasp the file handle with the thumb and index finger of the right hand, resting the end of the handle lightly against the base of the thumb. The end of the file is guided with the thumb and index finger of the left hand with the end of the thumb on top and the tip of the index finger beneath, figure 14-12.
9. Test the finished work with a straightedge or a combination square.

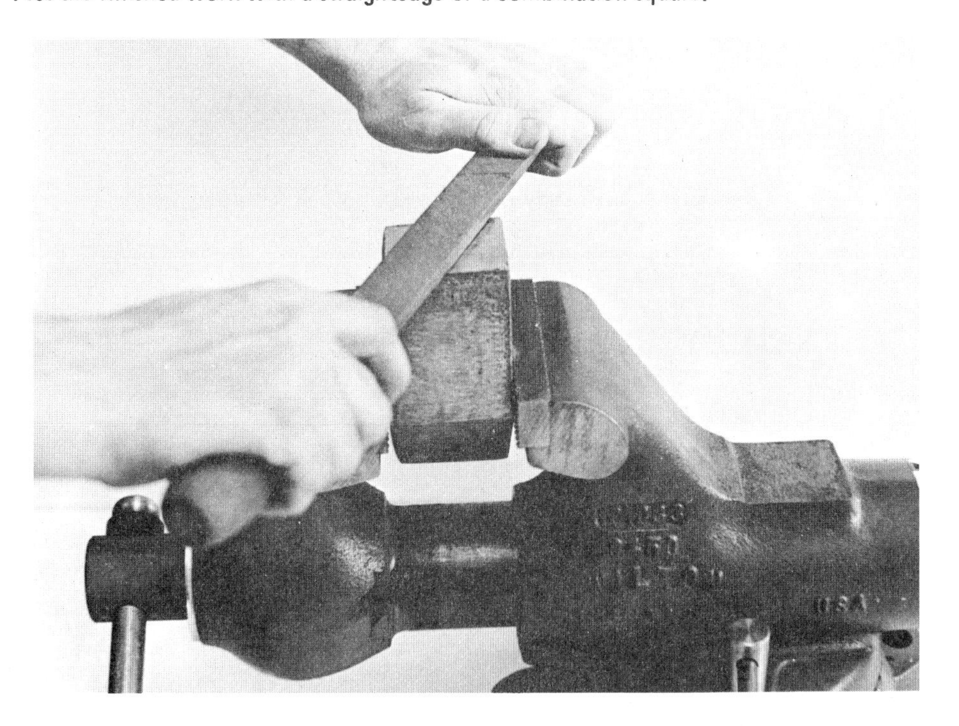

Figure 14-12 Finishing a steel surface.

HOW TO FILE CORNERS

A. *Outside Corners*

1. Clamp the work in a vise so that the layout line is above the vise jaws.
2. Rough file to the layout line by filing across the piece.
3. Reduce the corner with a series of corners until the required radius is obtained, figure 14-13.
4. Finish filing with a fine-cut file by following the rounded surface.

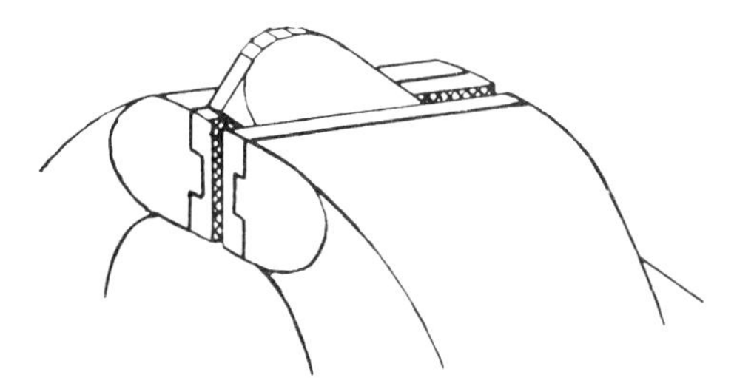

Figure 14-13 Rounding a corner.

B. *Inside Corners*

Select a file of the proper shape to suit the outline of the opening. Use a hand or pillar file with a safe edge for square corners and file across the edge. For curved corners, use a round or half-round file and give the file a slight twist on the corners.

HOW TO REMOVE BURRS AND SMOOTH JAGGED, SHARP EDGES

1. Select a single-cut file which has been broken in properly.
2. Hold the work in a vise or clamp it on a bench, depending on its size.
3. For short distances, file across the edge. When the edges are long, hold the file at a slight angle to the edge and push the file along the edge for the entire length of the sheet as shown in figure 14-14. A chamfered edge is produced in the same way.

When filing with one hand, place the forefinger instead of the thumb on top of the file in order to control the tool better.

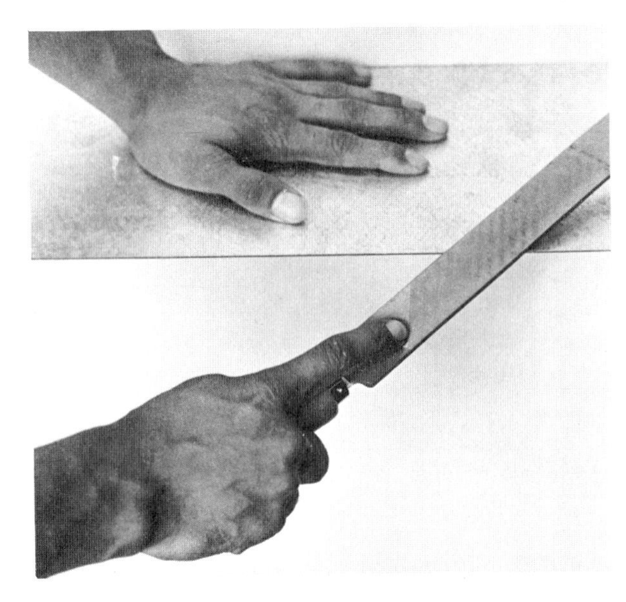

Figure 14-14 Removing a burr.

SUMMARY REVIEW

A. Place the answers to the following questions in the column to the right

1. List the parts of a file. 1. ______________________
   ____
   ____
   ____
   ____

2. What type of handle is considered the safest for use on the tang of a file? 2. ______________________

3. List the purposes for which a sheet metal worker uses a file. 3. ______________________
   ____
   ____
   ____

4. List four shapes of commonly used files. 4. ______________________
   ____
   ____
   ____

5. List the two general types of cuts for files. 5. ______________________
   ____

6. What type of file should be selected to 6. ______________________
   a. remove the metal surface rapidly
   b. to finish up a metal job? ______________________

7. What type of file should be selected to remove excess lead from a soldered joint? 7. ______________________

8. A square hole is cut in a duct, with jagged edges and corners. What type of file should be used to correct the cut? 8. ______________________

9. If a new file is to be used on a job, how should it be prepared? 9. ______________________

10. Why is pressure released on the back stroke in filing? 10. ______________________
    ____

B. Insert the correct word in each of the following sentences.

1. The file is made to cut on the __________ stroke only.
2. Always file the rough-cut edge of a thin piece of sheet metal in (a, an)____________ direction.
3. An uneven surface can be caused in filing by__________ the file.
4. On the forward cutting stroke,__________ is applied to the file.

C. Underline the correct word in each of the following sentences.

1. A file can be used for (prying, hammering, finishing) a steel surface.
2. A clogged file should be cleaned with a (cloth, steel block, file card).
3. Always (hammer, tap, force) a file handle on the tang.
4. When filing a surface, work should be tested for straightness with a (rule, square, tape).

UNIT 15 *PUNCHES*

OBJECTIVES

After studying this unit, the student should be able to

- Identify and describe the various types of punches.
- Select the proper type and size punch for the job.
- Use the punch correctly.

When holes are needed in sheet metal, they may be either punched or drilled. When punching holes, a tool called a punch is forced through the metal. Light sheet metal (24 gage) is more successfully punched than drilled since the drill is apt to leave burrs at the edge of the hole. Punching is also quicker than drilling. For heavier metal, drilling is more accurate and distorts the metal less than punching. Drilling is, therefore, preferable for heavy work or for any place where maximum accuracy and strength is required.

There are several kinds of punches: solid punches and lever punches for small diameter holes, hollow punches for large diameter holes, lever punches for heavier metals, and machine punches for production work.

BACKING FOR SOLID AND HOLLOW PUNCHES

When a hole is punched in a sheet metal with a solid or hollow punch, some sort of backing is necessary, This backing must be solid and at the same time give somewhat whenever force is applied.

A lead cake or the end grain of a block of wood may be used for backing. The hole must be cut without stretching the metal too much so that the hole is not too small after the metal has been flattened. The cross grain of a block of wood should not be used because it is soft and allows the metal to stretch before it is cut. Figure 15-1 shows the effects of good and bad backing practice. Note the distortion of the metal around the hole when punching is done on the cross grain of a block of wood.

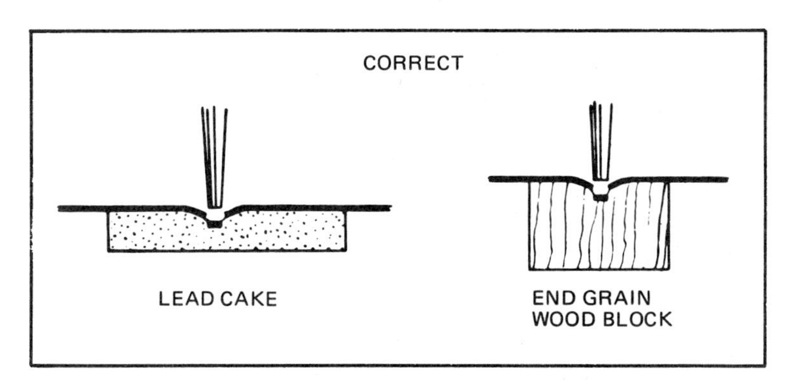

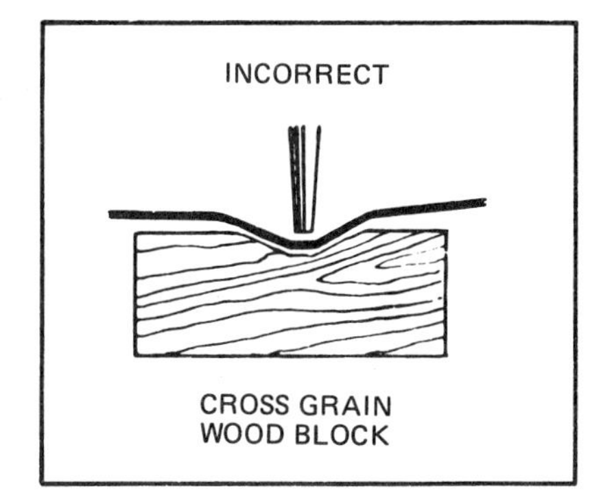

Figure 15-1 Using proper backing.

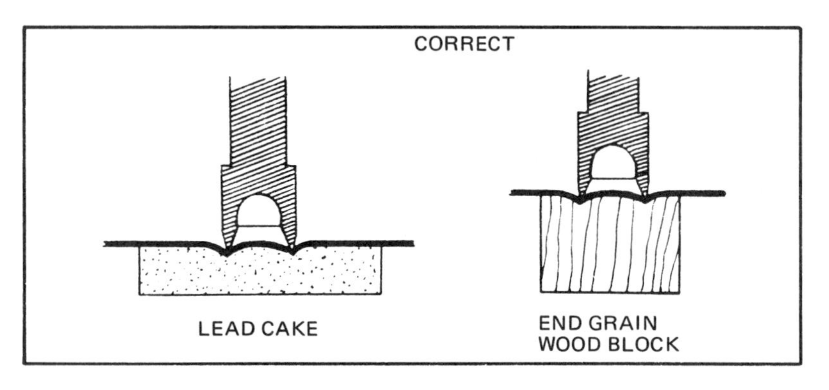

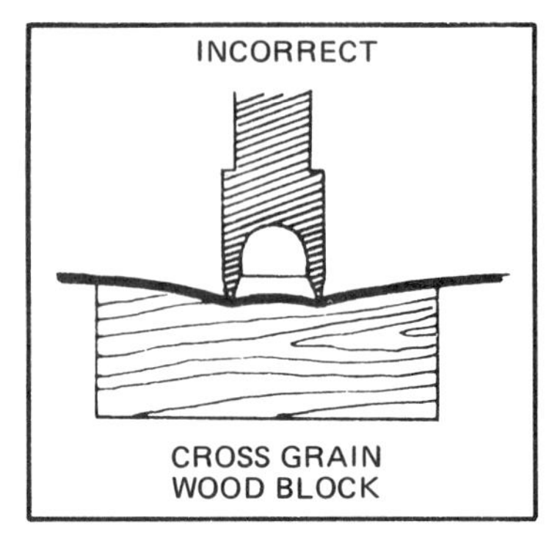

Figure 15-1 Using proper backing. (Cont'd)

A steel plate or an iron rail should not be used to back up a job being punched because it does not allow the end of the punch to pass through the sheet metal and thus ruins the punch.

After the holes are correctly punched, figure 15-2A, the metal is shaped around the hole. The metal around the hole is then flattened on a bench plate or rail with a mallet, figure 15-2B. If a hammer were used, the metal around the hole would be stretched. This would reduce the size of the hole and bulge the metal as shown in figure 15-2C.

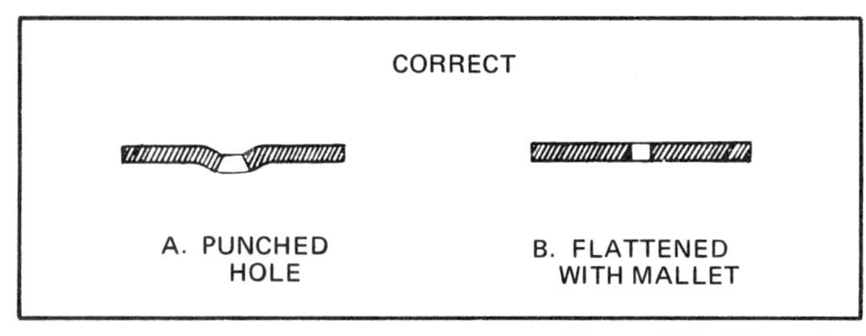

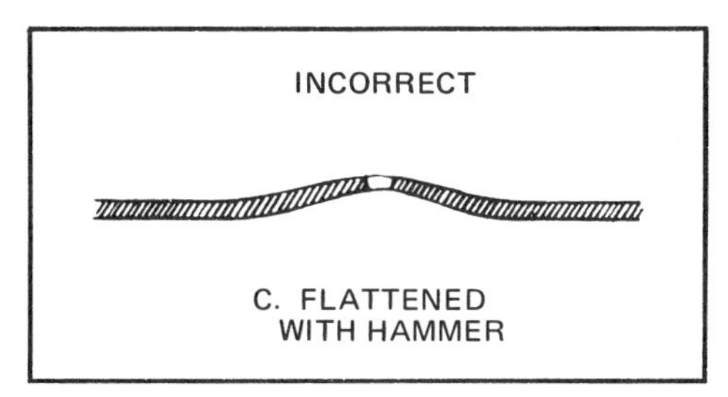

Figure 15-2 Flattening the metal.

SOLID PUNCHES

A solid punch is a hardened circular or octagonally shaped rod of tool steel with ground ends, figure 15-3. The circular punches are knurled so they can be held without slipping. Some solid punches have a center point to facilitate centering the punch.

Solid punches are used to punch holes from 3/32 to 1/2 inch in sheet metal of 24 gage and lighter. These punches come in several different sets.

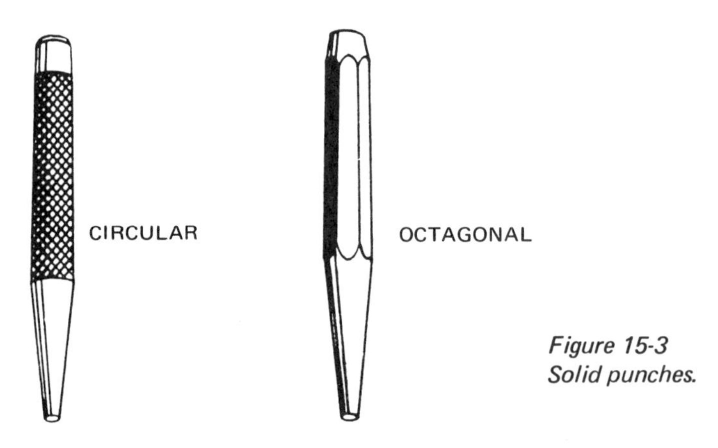

Figure 15-3
Solid punches.

HOW TO USE A SOLID PUNCH

1. Lightly mark the center of the hole to be punched with a prick punch.

If the head of the prick punch is mushroomed, it should be ground before using.

2. Place the material to be punched flat on a lead cake or the end grain of a block of wood.
3. Select a solid punch of the proper size and place it directly over the prick punch mark.

If the head of the solid punch is mushroomed, it should be ground before using.

4. Hold the solid punch in a vertical position and strike it a medium blow with a heavy hammer, figure 15-4.

Watch the end of the punch--not the head of the punch.

Figure 15-4 Using the solid punch.

5. Raise the solid punch and make sure that the punch is in the center of the prick punch mark. (This is not necessary if the punch has a center point.)
6. Return the solid punch to the impression previously made and continue striking sharp blows with the hammer until the metal is punched.

Usually one light and one sharp blow are sufficient if the solid punch is sharp, and the backing is on a solid foundation.

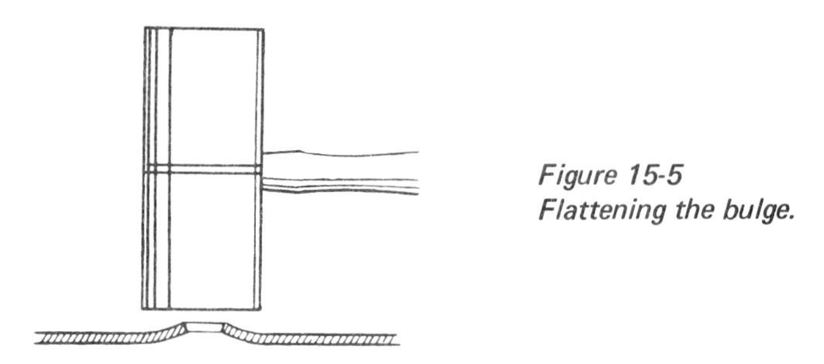

Figure 15-5
Flattening the bulge.

7. If the sheet is to be smooth on both sides, the sheet is placed on a plate with the bulged side up and flattened with a mallet, figure 15-5.

HOLLOW PUNCHES

Hollow punches are rods of tool steel having hollow ends with beveled cutting edges, figure 15-6. The shanks are knurled to facilitate gripping. Some hollow punches have a center pin to aid in centering the punch.

Hollow punches are used to punch larger holes than solid punches, and the sizes start approximately where the solid punch sizes end. Generally, the sizes vary from 1/4 to 4 inches in diameter, with 1/8-inch increments.

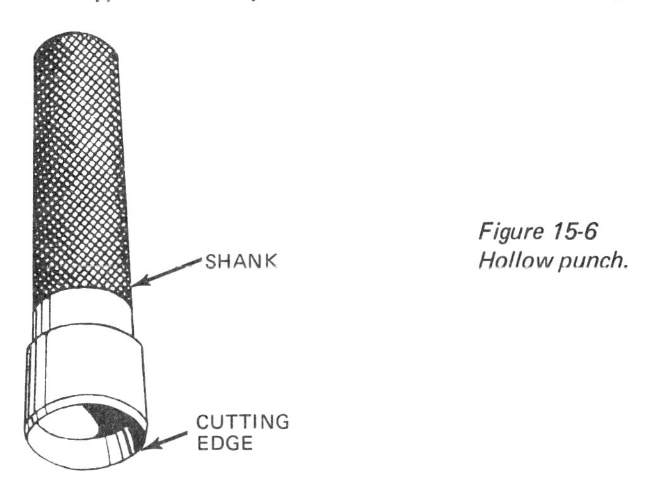

Figure 15-6
Hollow punch.

HOW TO USE A HOLLOW PUNCH

1. Mark the center of the hole to be punched with a prick punch.
2. Set the dividers for the radius of the hole to be punched.
3. With the prick punch mark as a center, scribe the circle on the metal with the dividers.
4. Select a hollow punch to correspond to the diameter of the hole.

If the head of the hollow punch is mushroomed, it should be ground before using.

5. Place the material flat on a lead cake or the end grain of a block of wood.
6. Hold the hollow punch in a vertical position directly over the scribed circle and strike it a medium blow with a heavy hammer, figure 15-7.

Watch the end of the hollow punch—not the head.

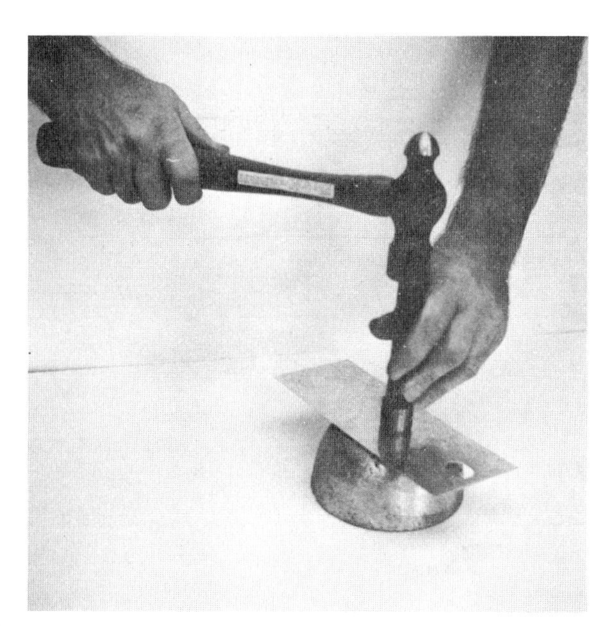

Figure 15-7 Using the hollow punch.

7. Raise the hollow punch and make sure that it is directly over the scribed circle.

If the hollow punch has a centering pin, it is not necessary to raise the punch.

8. Return the hollow punch to the impression previously made and continue striking sharp blows with the hammer until the metal is punched.

The number of blows required depends on the sharpness of the hollow punch, the weight of the hammer used, and the firmness of the foundation that the backing rests on.

9. If the sheet is to be smooth on both sides, the sheet is placed on a plate with the bulged side up and flattened with a mallet.

LEVER PUNCHES

A lever punch consists of a die and a punch moved by levers. The punch forces the metal through the die, leaving a clean hole. The die acts as a backing instead of the wood block or lead cake. Both the punch and the die are removable. Like solid punches, the lever punch is used for punching small holes in light sheet metal. It is quicker but can be used only when the holes are to be punched near the edge of a sheet. Specialized punches have been developed for other seaming methods such as the button punch, the clip punch, and the snap lock punch.

Types of Lever Punches

The punches and dies of the lever punch are made of hardened and ground tool steel. The punch has a shoulder which fits into a collar in the type of hand punch shown in figure 15-8A. This collar is threaded and is fastened to the punching lever. At the lower end of the punch is a center point which is used for locating the center of the hole to be punched. The die is threaded and has a slot in the lower end so that it can be removed easily.

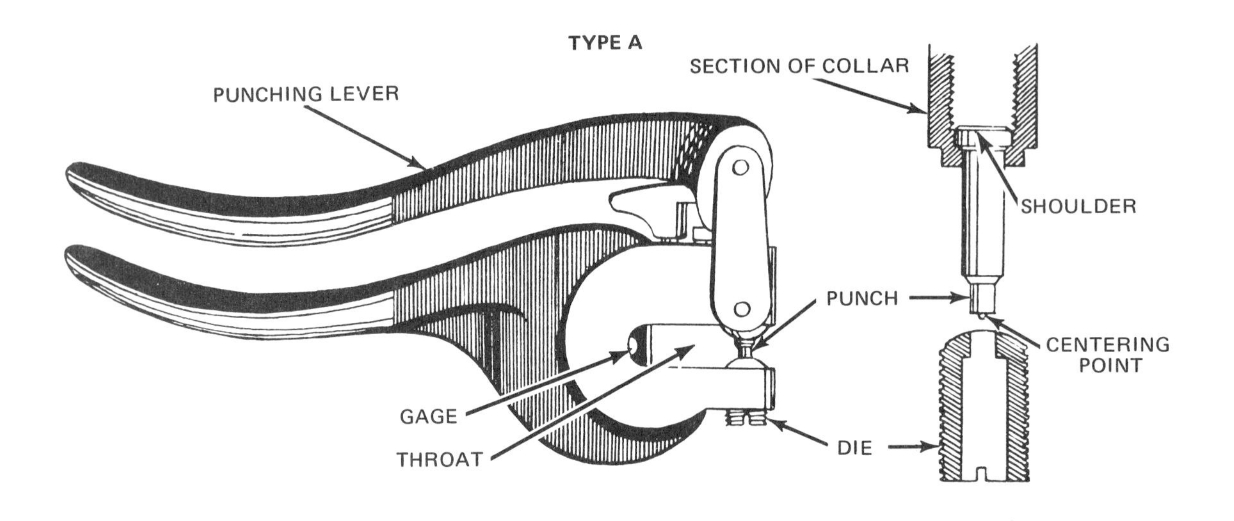

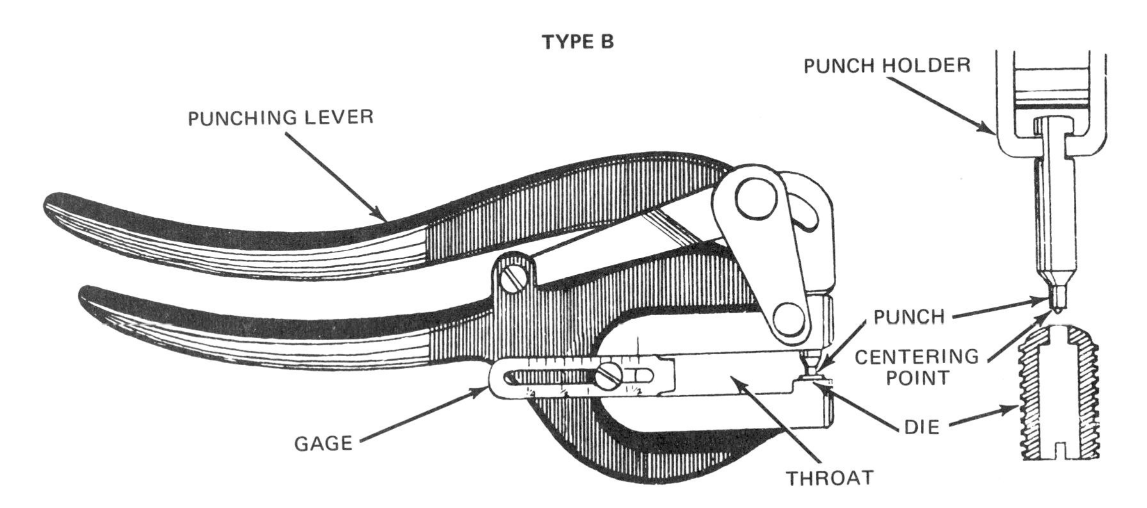

Figure 15-8 Two types of lever punches.

Clearance is provided inside the die to allow the punched material to fall freely. The depth of the throat governs the distance from the edge that a hole can be punched. The lever punch shown in figure 15-8B has the same general construction, but the levers are arranged differently and the dies and punch are changed in a different manner. The punch is moved by the punching lever and another lever called the punch holder. This punch holder has flanges which fit into the recesses of the punch.

Both punches leave clean, accurate holes and are used for punching rivet holes and bolt holes. With the hand punch shown in figure 15-8A, holes can be punched in the center of a 2 1/2-inch circle; with the hand punch shown in figure 15-8B, holes can be punched in the center of a 3-inch circle. There are other small lever punches available with a 3-inch throat capable of punching a hole in the center of a 6-inch circle or square.

The capacity of the punches shown is 9/32 inch hole through 16-gage mild steel. Larger holes and heavier gage metal can be punched with lever punches with longer lever arms.

HOW TO CHANGE PUNCH AND DIE: TYPE A

1. Remove the die with a screwdriver as shown in figure 15-9A.
2. Open the punch and hold it in the position shown in figure 15-9B.
3. Remove the threaded collar with a small wrench as shown.

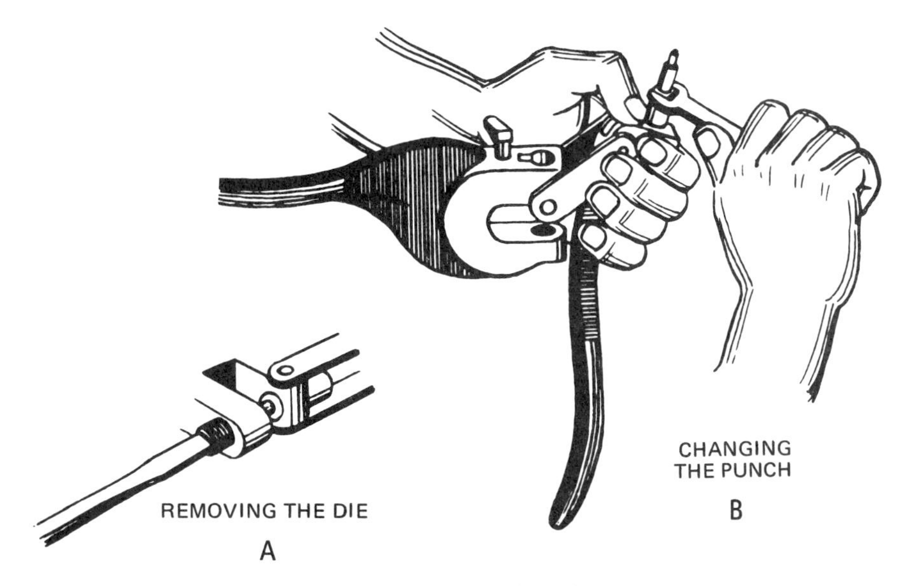

Figure 15-9 Changing the punch and die: Type A.

4. Slide the punch out of the collar.
5. Insert a punch of the required size in the collar.
6. Replace the threaded collar with the small wrench.
7. Return the levers to their normal position.
8. Insert a die of the required size.
9. Adjust the die with a screwdriver until the end of the punch enters the die about 1/32 to 1/16 inch when the levers are closed.

HOW TO CHANGE PUNCH AND DIE: TYPE B

1. Remove the die with a screwdriver.
2. Remove the knurled, slotted screw that holds the punch holder in place with a screwdriver, figure 15-10.
3. Raise the punching lever.
4. Slide the punch holder back far enough to release the punch.

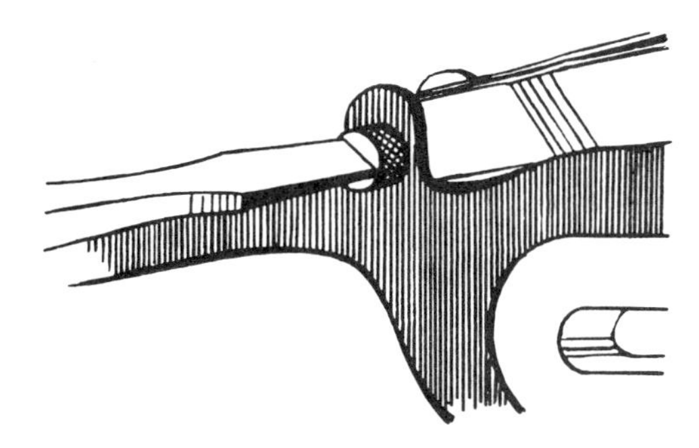

Figure 15-10 Removing the knurled screw.

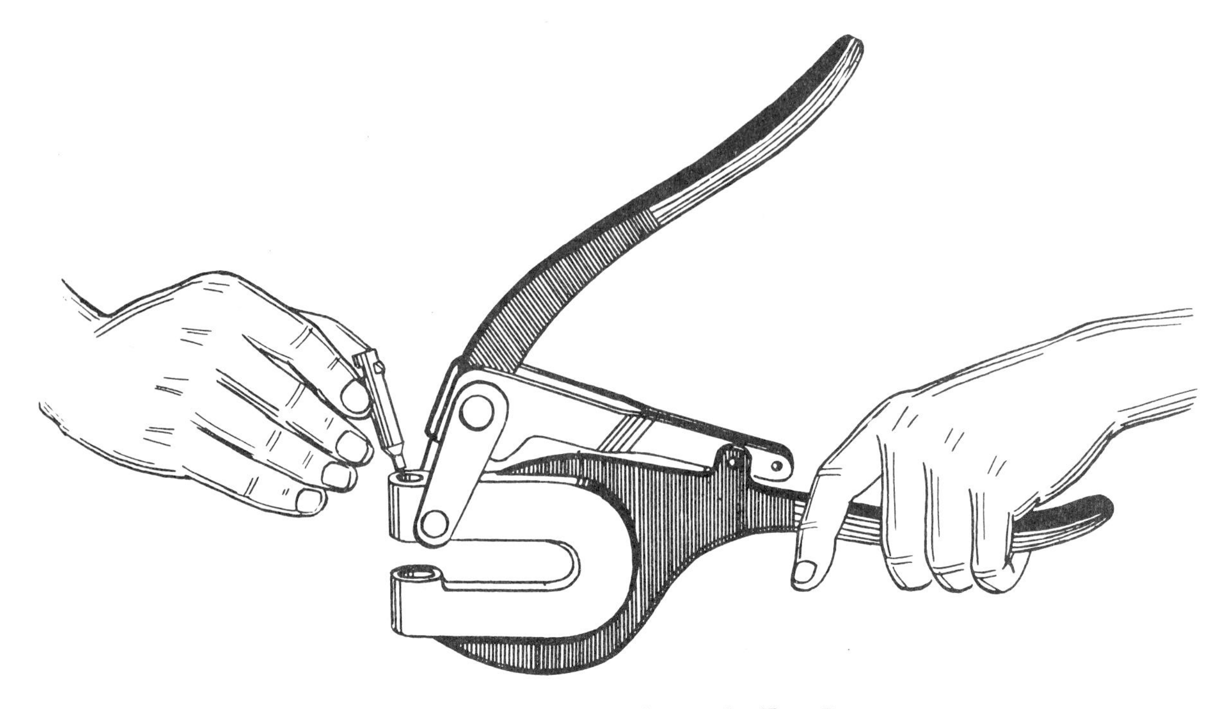

Figure 15-11 Removing the punch: Type B.

5. Hold the hand punch as shown in figure 15-11 and remove the punch.
6. Insert a punch of the required size with the recesses parallel to the flanges of the punch holder.
7. Bring the levers together so that the flanges on the punch holder slide into the grooves of the punch, figure 15-12.
8. Replace the knurled, slotted screw at the center of the punch.
9. Insert a die of the proper size.
10. Adjust the die until the end of the punch enters the die about 1/32 to 1/16 inch when the levers are closed.

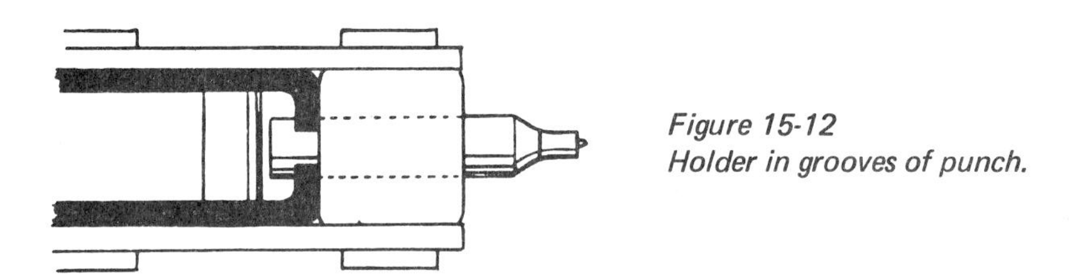

Figure 15-12
Holder in grooves of punch.

HOW TO USE THE LEVER PUNCH

1. Lay out the holes in the required places on the pattern and mark the centers with a prick punch or center punch.

Use a prick punch for metal of 24 gage and lighter and a center punch for metal 22 gage and heavier.

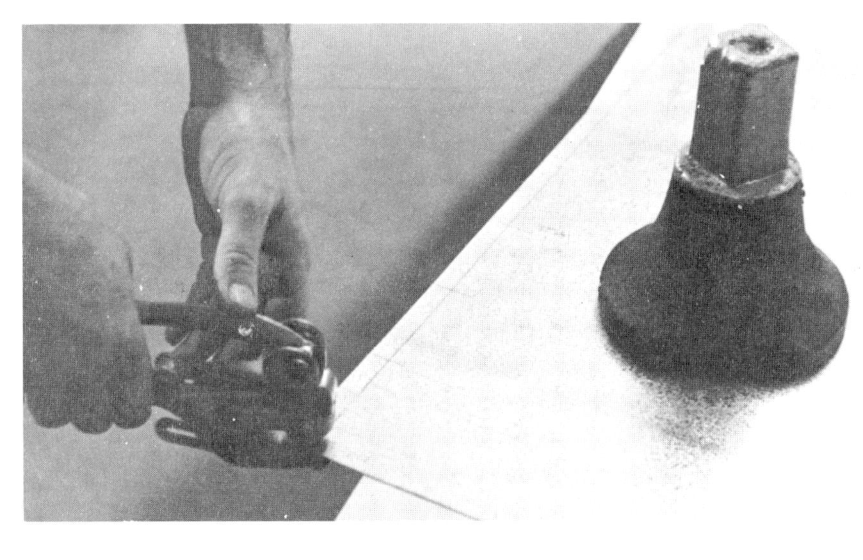

Figure 15-13 Using the hand punch.

2. Make sure that the punch of the correct size is in the lever punch. If it is not, change the punch and die.
3. Lay the sheet on the bench with the marked holes extending beyond the edge of the bench.

CAUTION When punching medium-sized pieces, weights are placed on the sheet to hold it in place.

4. Place the centering point of the punch in the center mark on the sheet, figure 15-13.
5. Hold the throat and the lower lever in a horizontal position.
6. Punch the hole by pressing the upper lever down.

When very small pieces are to be punched, it is sometimes more convenient to fasten the hand punch in a vise and feed the metal to the punch. Figure 15-14 shows the punch held in a bench vise and a small piece of metal in position for punching.

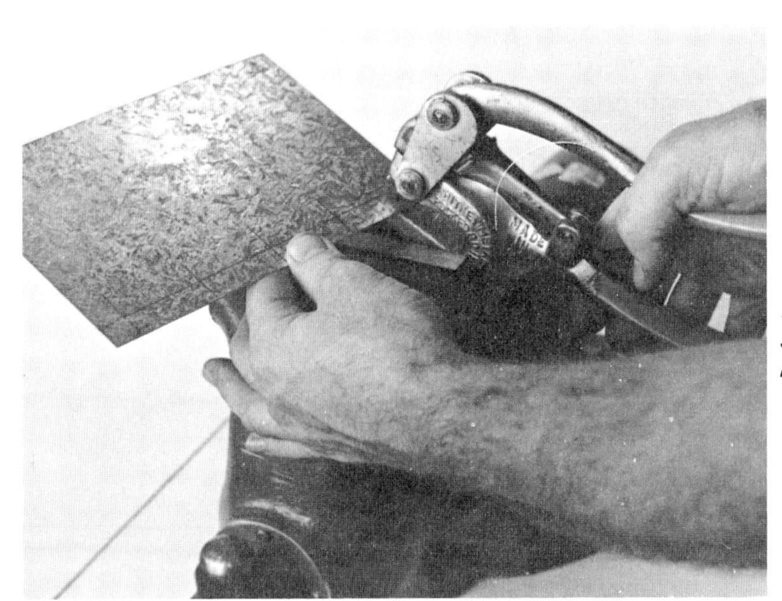

Figure 15-14 Punching a small piece with the punch held in a vise.

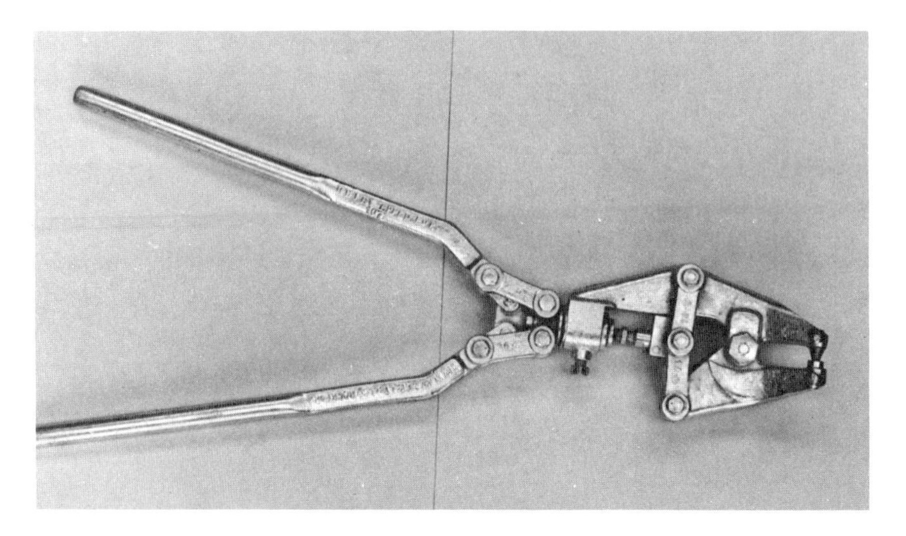

Figure 15-15 Button punch.

THE BUTTON PUNCH

The button punch, figure 15-15, is used to indent several thicknesses of metal together to provide a secure fastening in place of a hole and riveting. It is commonly used on standing seams on roofs to make a watertight fastening. The head of the punch can be revolved to any desired position independent of the handles. Jaws are set at a 30-degree angle to permit close corner work and insure visibility. The capacity of the punch is 4 thicknesses of 24-gage mild steel. Its throat is 1-3/4 inches deep.

HOW TO USE THE BUTTON PUNCH

1. Mark the locations of the buttons on the seam with prick punch indentations 2 inches apart on centers.
2. Place the punch and die over the indentations.
3. Squeeze the handles firmly.

THE CLIP PUNCH

The clip punch, figure 15-16, is used to join ductwork seams together in heating and air conditioning work. Flat-nosed jaws permit easy access to cramped places. The capacity of the punch is 3 thicknesses of 24-gage mild steel. The throat depth is 1 inch. The punch itself turns down an ear which is later hemmed tight with a hammer.

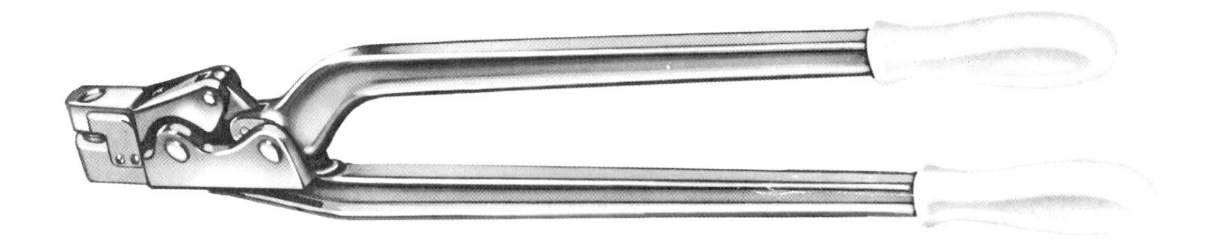

Figure 15-16 Clip punch. (Courtesy of Roper Whitney, Inc.)

HOW TO USE THE CLIP PUNCH

1. Mark the location of the clips with a scratch awl along the seam–2 inches on centers.
2. Place the punch over the desired location.
3. Squeeze the handles until punch pierces metal and turns an ear down. The seam is finished by hammering the ears down flat.

THE SNAP LOCK PUNCH

The snap lock punch, figure 15-17, is an extremely useful tool for duct and furnace installation. The capacity of this tool is 20-gage mild steel. It forms lugs or ears that lock firmly on the seamed edge of a joining piece.

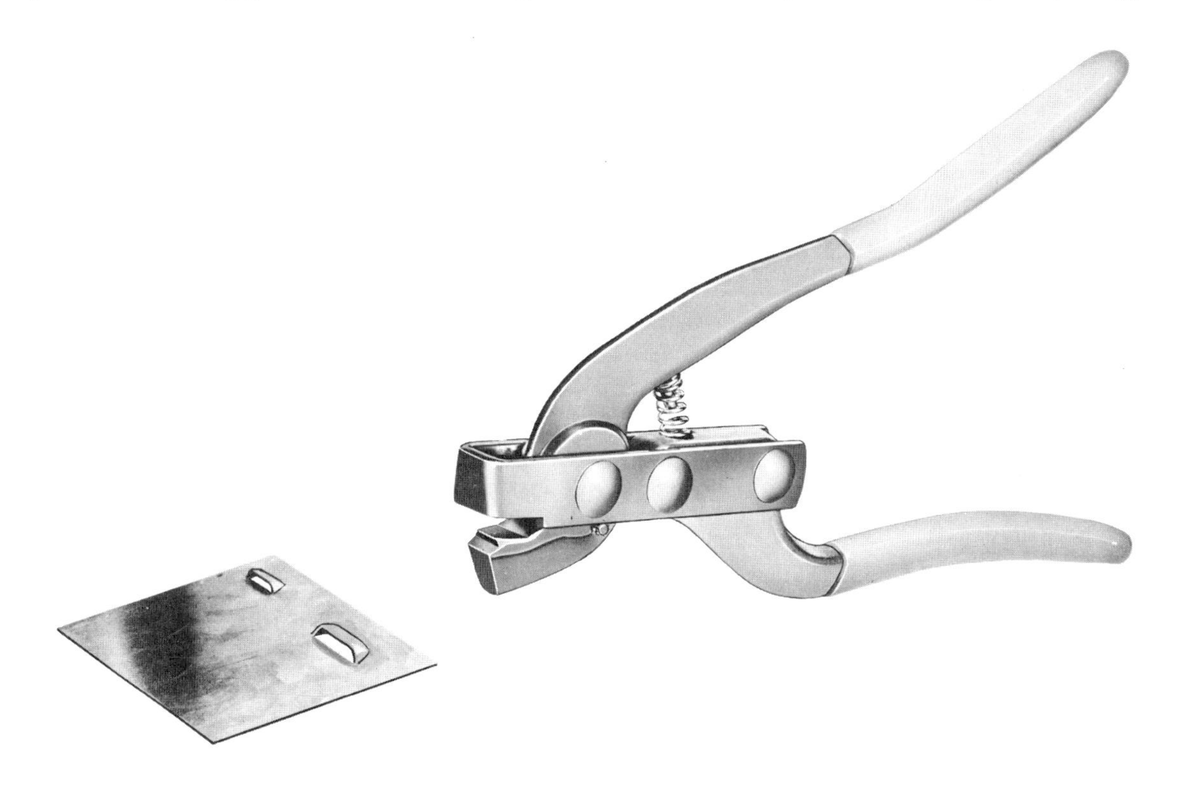

Figure 15-17 Snap lock punch. (Courtesy of Roper Whitney, Inc.)

HOW TO USE THE SNAP LOCK PUNCH

1. Mark the location of the locks with a scratch awl along the flanged seam edge–2 inches on centers.
2. Place the jaws on the flanged edge of the marked location.
3. Squeeze the handles together firmly to form a lug. The seam is finished by forcing the snap lock edge into the pocket.

SUMMARY REVIEW

A. Place the answers to the following questions in the column to the right.

1. List the kinds of punches used in the metal trade. 1. ______________________

____

____

____

2. Why is it preferable to punch a hole in light metal rather than drill it? 2. ______________

3. Why is it necessary to back up a piece of sheet metal when punching a hole with a solid punch? 3. ______________

4. List two materials that make good backing blocks for hole punching in galvanized iron. 4. ______________ ______________

5. Which is softer, the cross grain or the end grain of a piece of wood? 5. ______________

6. List the two essential parts of a lever punch. 6. ______________ ______________

7. Why are punches and dies removable from the lever punch? 7. ______________

8. What governs the distance from the edge that a hole can be punched with a lever punch? 8. ______________

9. What, in general, is the difference between the two types of lever punches? 9. ______________

10. On what type of seam is the button punch commonly used? 10. ______________

B. Insert the correct word in each of the following sentences.

1. Metals distorted by punching can be corrected with (a, an) ____________ .
2. Punches are __________ so they can be held without slipping.
3. The __________ of the button punch can be revolved to any desired position.
4. The __________ punch turns down an ear on the seam.
5. The ear formed by the punch is later hemmed tight with (a, an) .

C. Underline the correct word in each of the following sentences.

1. If a ball peen hammer is used to correct distortion from punching a hole, the metal will (shrink, stretch, bulge, mark.)
2. When punching a hole for a 5/16-18 bolt in 24-gage sheet metal, a (1/4, 3/8, 1/2-inch) hollow punch is used.
3. In making a seam for round pipe the (button, clip, snap lock) punch is used.
4. The lever punch forces metal through a die leaving a (burred, distorted, clean) hole.
5. The lever punch is used extensively for punching (rivet, screw, nail) holes.

UNIT 16 HAND AND BREAST DRILLS

OBJECTIVES

After studying this unit, the student should be able to

- List the parts of a hand drill.
- State the capacity of the hand and breast drills.
- Select the proper drill for the job and use it correctly.

The hand drill is used by the sheet metal worker for drilling holes in light gage metal when a power tool is not applicable. When installing aluminum gutters, this type of drill is useful for drilling rivet holes in fastening lengths together. The two general types of drills are the small hand drill and the larger breast drill.

TWIST DRILL BITS

Twist drill bits are made of carbon steel or high speed steel. They can be used in hand or breast drills as long as they have a straight shank. Twist drill bits are available for use in the hand or breast drill in fractional sizes ranging from 1/64 inch to 1/2 inch in intervals of 1/64 inch. Sizes are marked on the shank.

Twist drills should have sharp cutting edges (lips) centered about the fluted body.

HOW TO SHARPEN A DRILL BIT

1. Hold the shank of the drill bit in the right hand and the body in the left, figure 16-1.

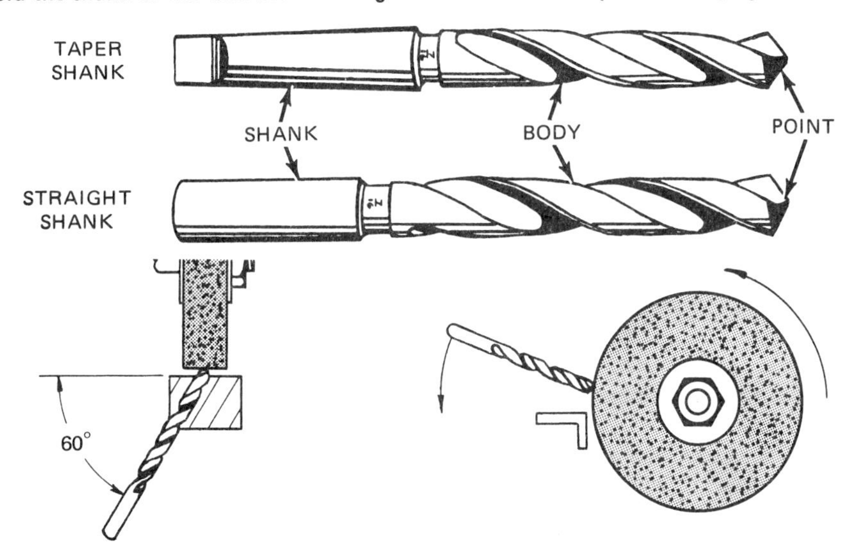

Figure 16-1 Sharpening a drill bit.

2. Stand so that the bit is at an angle of approximately 60 degrees to the face of the grinding wheel.
3. Touch one cutting edge to the wheel in a horizontal position and pivot the shank downward with the right hand while rotating the bit very slightly.

Cool the bit frequently in cold water to avoid burning it during sharpening.

4. Repeat step 3 until half the angle is ground.
5. Turn the drill bit to the other cutting edge and repeat steps 1 to 4.

THE HAND DRILL

Hand drills are used to drill holes 1/4 inch or less in diameter. The maximum chuck capacity for the hand drill is 1/4 inch. For larger size drill bits, the breast drill is used.

Many manufacturers provide a hollow handle with a screw top for the storage of drills. An assortment of drill bits ranging in sizes from 1/16 to 11/64 inch in diameter are included.

HOW TO INSERT AND REMOVE TWIST DRILL BITS

1. Open the chuck slightly more than the diameter of the drill bit and insert the bit. Tighten the chuck, by pushing forward on the crank with the right hand, while holding the chuck shell with the left hand, figure 16-2.
2. To remove the bit, hold the chuck shell with the left hand and turn the crank backward with the right hand as shown by the arrow, figure 16-3.

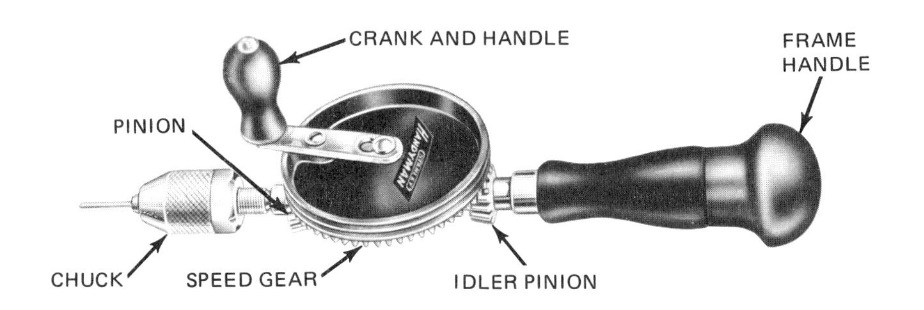

Figure 16-2 Hand drill. (Courtesy of Stanley Tools.)

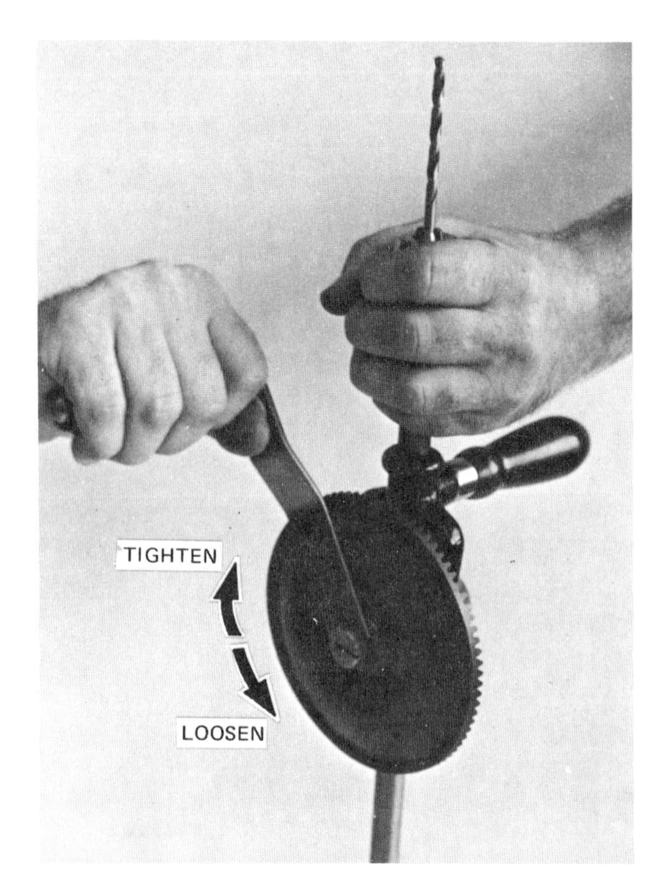

Figure 16-3 Loosening or tightening a drill bit in the chuck.

HOW TO DRILL HOLES WITH A HAND DRILL

1. Locate the holes to be drilled with intersecting lines. Make a mark on the metal with a center punch.
2. Insert the bit in the chuck. Make sure that it is fastened tightly and that it bottoms in the chuck.
3. Rest the point of the drill on the mark and rotate the crank and handle at a moderate speed. Use only enough pressure on the handle to cut the metal. Hold the drill steady so that the hole will not be drilled oversize and the bit will not break. The drill should be perpendicular to the surface of the work, figure 16-4.
4. When the metal is almost pierced, reduce the speed of drilling and the pressure applied.

This should be done when drilling through holes in wood so that binding and splintering will not occur as the drill breaks through. When drilling metal, the change in speed and pressure prevents breaking the drill bit. Do not drill beyond the length of the twist on the drill because the chips cannot be cleared from the hole.

Figure 16-4 Drilling a vertical hole.

5. When drilling horizontal holes, use the drill as illustrated in figure 16-5A. Position the drill for squareness of drilling. The gear to which the crank handle is attached is held on a vertical plane to the right of the drill body. Better control of the drill is thereby possible. It is sometimes desirable to hold the drill by the side handle and press the body against the frame handle like a breast drill, figure 16-5B.

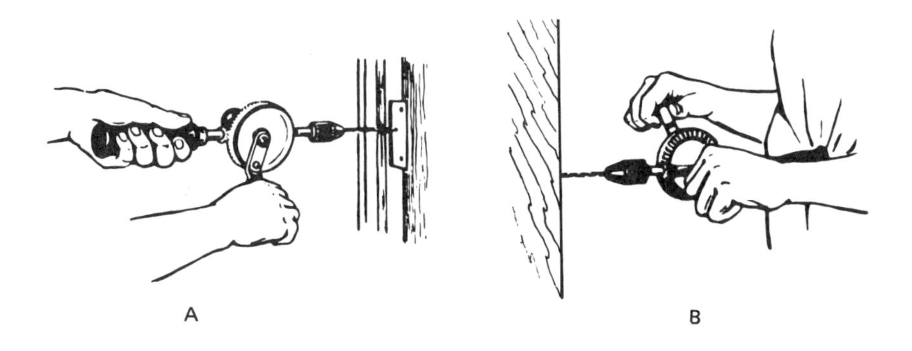

Figure 16-5 Drilling a horizontal hole.

BREAST DRILL

A breast drill is a larger and more strongly built version of the hand drill, figure 16-6.

This type of drill usually has a chuck capacity for drill bits ranging up to 1/2 inch in diameter. As with the hand drill, straight-shank drills are used in the chuck.

The breast drill is fitted with a plate instead of a handle. Pressure is applied to the plate by the operator's chest or abdomen for feeding the drill. Most types provide for two speed changes.

A speed change is made by shifting and engaging the driving wheel spindle in one of two driving gears that are incorporated in the construction of the drill. When engaged in the larger gear, a comparatively high speed of spindle revolution is produced for using small size bits. It is better however, to use the regular hand drill for drilling holes less than 1/4 inch in diameter. The smaller driving gear transmits greater power but less speed to the spindle and is preferred for drilling large holes and hard materials.

Figure 16-6 Breast drill.

HOW TO USE A BREAST DRILL

1. Insert the drill in the chuck as instructed for the hand drill.
2. Locate and mark the hole with a scratch awl on wood or with a center punch on metal.
3. Locate the drill point on the mark. With the chest resting on the plate, drill by turning the crank and handle in a clockwise direction, figure 16-7.

Follow the same precautions as described for using the hand drill. Only a minimum of pressure should be applied to the plate. When drilling holes in sheet metal, it is always desirable to back up the drill area with a block of wood, as shown, to prevent distortion.

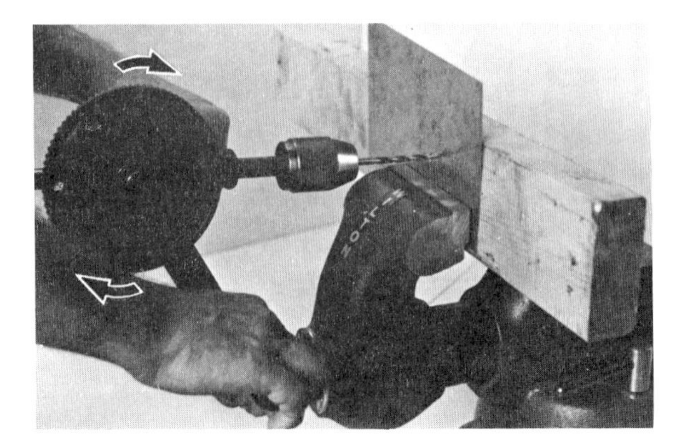

Figure 16-7
Using the breast drill.

SUMMARY REVIEW

A. Place the answers to the following questions in the column to the right.

1. For what kind of metal is the sheet metal worker most likely to use the hand drill? 1. ______
2. What size drill bit and what type of drill should be selected for drilling holes in 26-gage galvanized iron for 1 1/4-pound rivets (with diameter of .120)? 2. ______ ______
3. What is the maximum drill bit chuck size in the hand drill? 3. ______
4. List the parts of a hand drill. 4. ______ ______ ______ ______ ______ ______ ______
5. In what direction should the crank be turned to tighten the chuck? 5. ______
6. Why is a prick punch mark made at the center location of a hole? 6. ______
7. What will happen if a drill bit is allowed to wobble while turning? 7. ______
8. In what position should the drill be held in relation to the surface of the work? 8. ______

B. Insert the correct word in each of the following sentences.

1. If it is necessary to drill a 5/16-inch hole in a 1/4-inch plate without the power drill, the worker would use (a, an) ______ .
2. The maximum chuck capacity of a breast drill is ______ inch.
3. Pressure is applied to a breast drill by the ______ or ______ .
4. In drilling small holes in plate stock, the ______ speed is used.

C. Underline the correct word in each of the following sentences.

1. The chuck on a breast drill holds a (tapered shank, straight shank, square shank) bit only.
2. To drill a 3/16-inch hole in stainless steel, a (high speed, low speed) should be used.
3. In order to prevent distortion in drilling, the work should be backed up with a (steel plate, lead block, wood block).
4. The chuck capacity of a hand drill is (3/16, 1/4, 5/16) inch.

UNIT 17 *NOTCHERS, SEAMERS, AND CRIMPERS*

OBJECTIVES

After studying this unit, the student should be able to

- List the uses of special tools.
- Select the proper tool for the job.
- Use the selected tool properly.

There are different kinds of hand tools designed for use by the sheet metal worker to produce a specific job. The hand seamer, hand notcher, dovetail notcher, and pipe crimper are good examples. These hand tools are timesaving, portable, and efficient devices on the job. They are used on 22-gage metals and lighter stock.

HAND NOTCHER

The hand notcher, figure 17-1, is a lightweight tool designed for fast and accurate work. The hook design of the jaws permits notching to the exact desired depth without slippage and with a minimum of effort. A spring action returns the jaws to the open position.

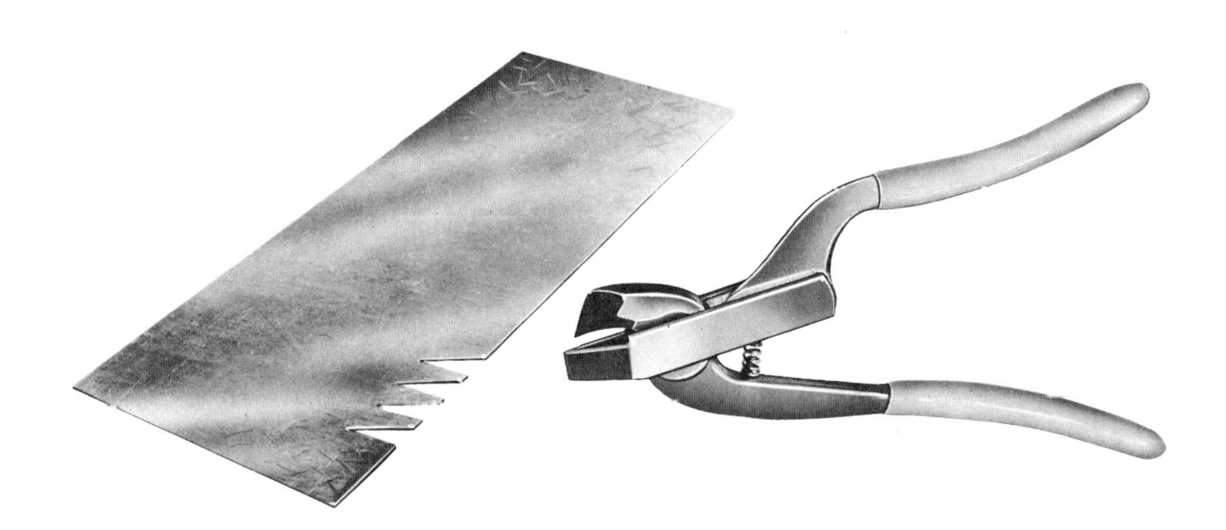

Figure 17-1 Hand notcher. (Courtesy of Roper Whitney, Inc.)

HOW TO USE THE HAND NOTCHER

1. Mark the depth of the notch with a gage line along the edge of the stock. Space by sight and avoid seams.
2. Place the hook of the notcher over the edge of the metal to the depth line.
3. Squeeze the handles together until the punched metal falls out.

THE DOVETAIL NOTCHER

The dovetail notcher, figure 17-2, is used on the ends of pipe, collars, and Y-joints. It is a lightweight, efficient small tool for dovetailing. It has the advantage of cutting the notch and folding the tab down square in one easy motion. The size of the dovetail is shown in figure 17-2.

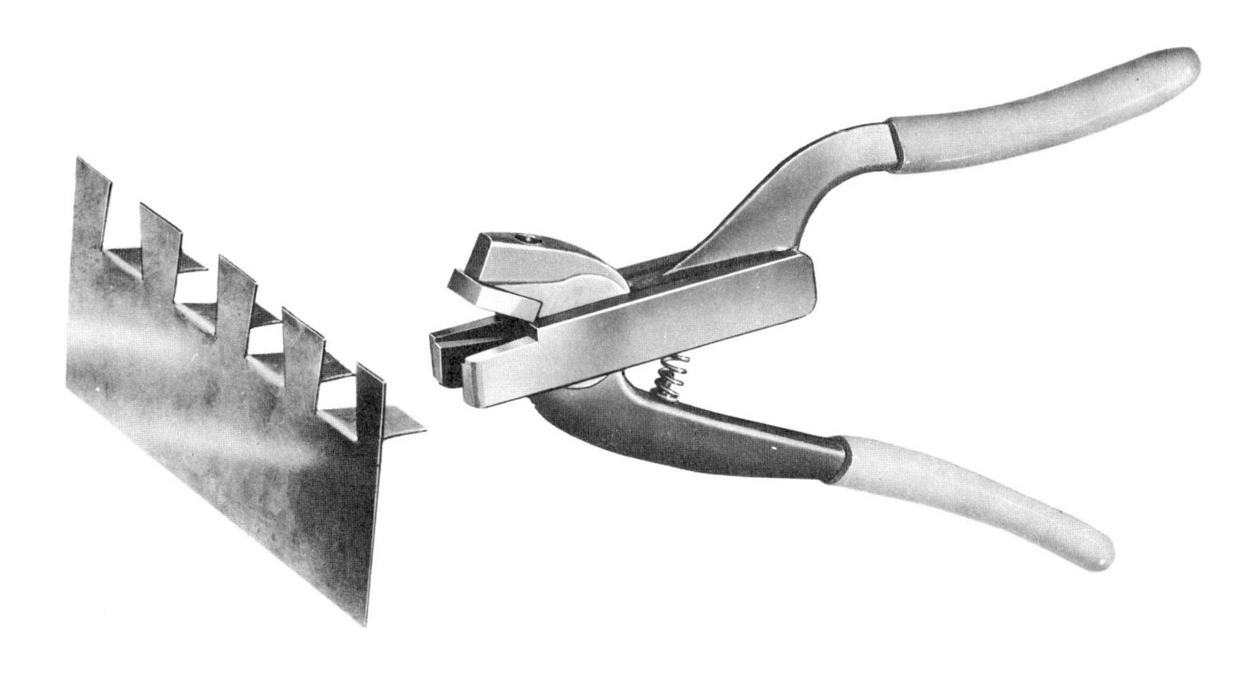

Figure 17-2 Dovetail notcher. (Courtesy of Roper Whitney, Inc.)

HOW TO USE THE DOVETAIL NOTCHER

1. Place the dovetail notcher over the edge of the stock to the full depth of the throat.
2. Squeeze the handles together to form a dovetail.
3. Space the notches by sight and avoid seams.

THE HAND SEAMER

The hand seamer, figure 17-3, is used extensively on heating and air conditioning ductwork. It is made of forged steel with 3 1/2-inch wide jaws. Some hand seamers are made with adjustable screws for setting the depth of the metal into the jaws. They bend stock when a hand brake or folder cannot be used.

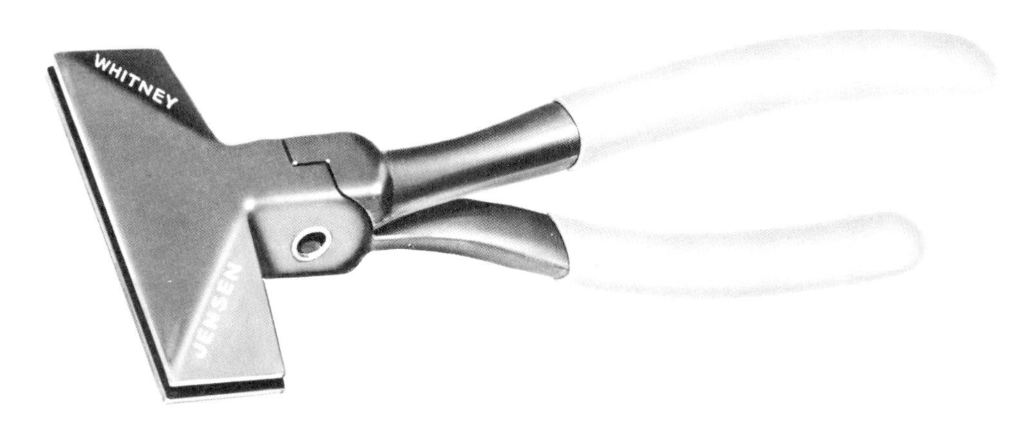

Figure 17-3 Hand seamer. (Courtesy of Roper Whitney, Inc.)

THE HAND SEAMER

1. Lay out the bend line on the metal.
2. Insert the metal between the jaws up to the bend line and squeeze the handles tight.
3. Hold the surface of the stock flat with the left hand and raise the seamer, figure 17-4.
4. If the stock to be worked is longer than the jaws of the seamer, bend small portions in succession.

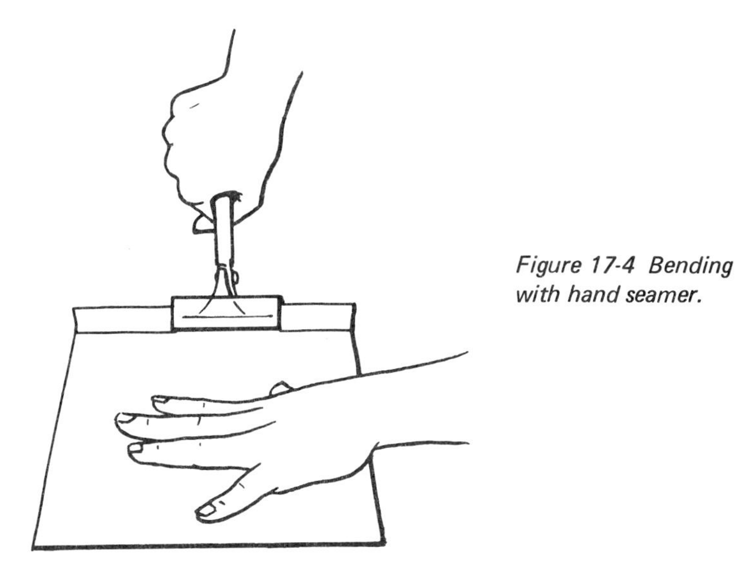

Figure 17-4 Bending with hand seamer.

THE HAND PIPE CRIMPER

The hand pipe crimper, figure 17-5, is designed to flute the edges of round pipe for nesting purposes. In gutter work, downspouts can be crimped on the edges regardless of the shape used to make nesting possible. Edges may be crimped along the entire perimeter, except where seams are made. Short crimps are used for tight nesting.

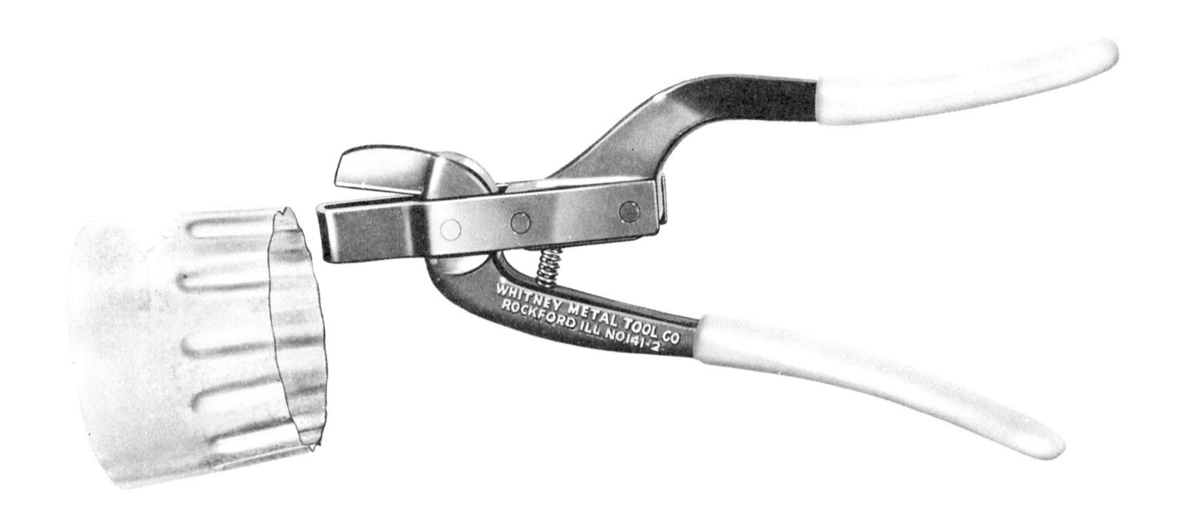

Figure 17-5 Pipe crimper. (Courtesy of Roper Whitney, Inc.)

HOW TO USE THE HAND PIPE CRIMPER

1. Place the hand pipe crimper over the edge of the pipe to the full depth of the throat.
2. Squeeze the handles until the crimp appears.
3. Regulate the depth of the crimp by hand pressure on the handles.
4. Space by sight around the pipe perimeter and avoid seams.

SUMMARY REVIEW

A. Place the answers to the following questions in the column to the right.

1. List the three advantages of a specialized hand tool. 1. ______________________
   ____
   ____
2. List three jobs in which a dovetail notcher is used. 2. ______________________
   ____
   ____
3. On a gutter job, 2 pieces of 4-inch round downspout are to be fitted together. How is the fitting done? 3. ______________________
4. When using the hand notcher or crimper, is it necessary to lay out each position? 4. ______________________
5. When is a shortened crimp used? 5. ______________________

B. Insert the correct word in each of the following sentences.

1. The hook design of the jaws of a hand notcher permits_______ to the exact desired_______ .
2. The dovetail notcher cuts the notch and _______ the tab down in one operation.

C. Underline the correct word in each of the following sentences.

1. A hand seamer is used for (breaking, twisting, bending) sheet metal.
2. If the bend is longer than the hand seamer jaws, bend (corners, centers, small portions).
3. When using the crimper or notcher, avoid the (edges, seams, perimeter) of a pipe.

UNIT 18 STAKES, PLATES, AND DOLLIES

OBJECTIVES

After studying this unit, the student should be able to

- Identify the various stakes used by the sheet metal worker.
- Describe the types of forming that can be done on the stakes.
- Select the proper stake for the job and use it correctly.

Sheet metal can be formed by bending or forming it over different anvils known as stakes. These stakes vary in shape, depending upon the type of work for which they are designed. Square, round, and conical work can be formed on the stakes, and edges and seams can be finished on them. Stakes are used when a suitable machine is not available or readily adaptable to the work. The sheet metal worker should know the many different shapes of stakes available so that he is able to select the proper stake for the shape and size of the job.

CARE OF STAKES

The sheet metal worker regards stakes as essential equipment and uses them as carefully as he would a delicate machine. Many of the stakes have hardened faces, but they should not be used for supporting jobs which are to be chiseled or punched. When chiseling or punching stock, use the common iron stakes for support.

To prevent marring the surfaces and the edges of the stakes, always return them to the place provided for them. A suitable rack for the stakes can be made by making a panel with the shapes and names of the stakes painted on it and fastening it securely to the wall. The stakes can be supported on the panel by hooks or projecting bolts.

BENCH PLATES

A bench plate or stake holder is used to hold the stakes while they are being used. The bench plate is a cast iron plate fastened to a bench, figure 18-1. Tapered holes of different sizes in the plate support the stakes. The smaller holes are used for bench shears.

The revolving stake holder can be mounted in any position desired by means of the clamp. Metal cabinet stands are also available for bench plates and individual stake mounting.

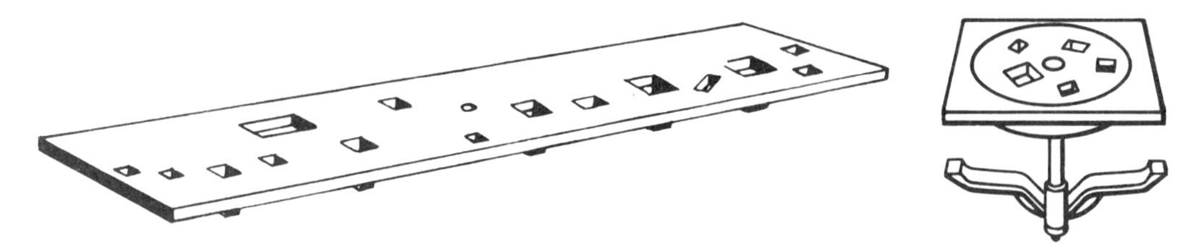

Figure 18-1 Bench plates.

IRON AND STEEL STAKES

Stakes come in a variety of sizes and shapes. Stakes have shanks which fit into the holes in the bench plate. The work is done on the heads or the horns of the stakes which are machined, polished and, in some cases, hardened. The stakes described below are shown in figure 18-2.

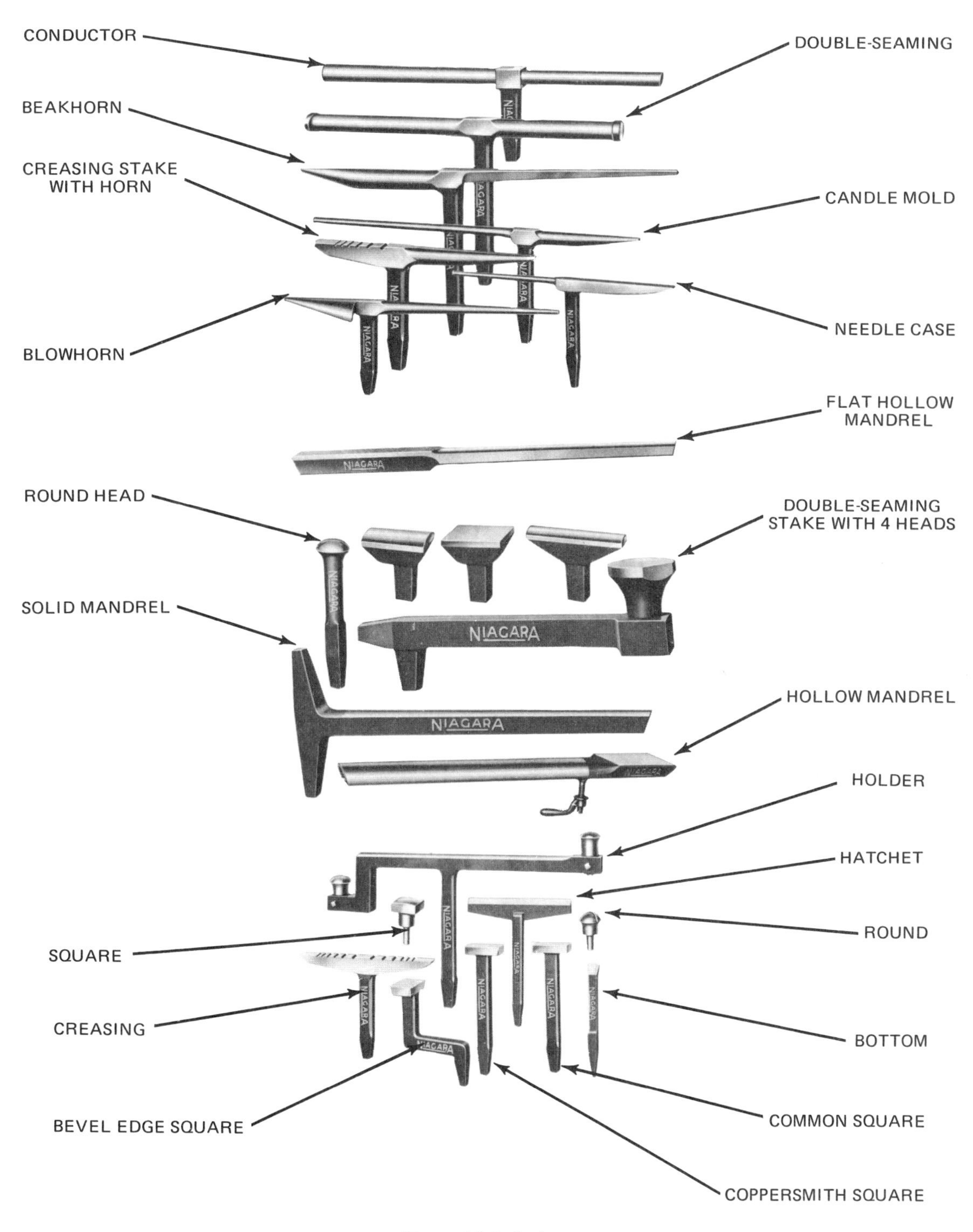

Figure 18-2 Stakes.

Square stakes have flat, square-shaped heads and are used for general work. There are three kinds: the *copper-smith stake;* which has one end rounded; the *bevel edge square stake,* which is offset; and the *common square stake.* Some of the edges are beveled so that they can be used for a greater variety of jobs.

The *hatchet stake* is a sharp, straight stake with a hardened edge. It is used for making sharp, straight bends and for folding and bending edges. The hatchet stake may also support the work when flanging and dovetailing. The *creasing stake* has one horn tapered for forming conical work. The other horn has grooved slots for turning, wiring, and beading operations. The *bottom stake* has a fan-shaped end, which is used for dressing a burred or flanged circular bottom.

The *conductor stake* has two cylindrical horns of different diameters. It is used when forming, seaming, and riveting pieces and parts of pipes. The *needlecase stake* has a round, slender horn for small tubes and wire rings and a heavier horn with a rectangular cross-section for square work.

The *blowhorn stake* has two horns with different tapers. The apron end shapes abrupt tapers, and the slender, tapered end is used for slightly tapered jobs. The large horn of the *candlemold stake* is a general-purpose stake, and the small horn is used for tube forming or reshaping. The *beakhorn stake* has a round, tapered horn on one end and a square, tapered horn on the other end. This general-purpose stake is used for shaping round and square work and for riveting.

There are two types of *double-seaming stakes.* One type backs up small, cylindrical jobs when double seams are laid down. It has two horns with elongated heads. The other type is used for double-seaming large work and for riveting. It has four interchangeable heads and two shanks, either of which can be placed in the bench plate so that the stake can be positioned horizontally or vertically.

The *hollow mandrel stake* has a slot in which a bolt slides so that the stake can be clamped rigidly to the bench. Either the rounded or the flat end can be used for forming, seaming, and riveting. There are two sizes available with an overall length of 40 and 60 inches. The *solid mandrel stake* has a double shank so that either the round or the square edge can be used. This stake is used for forming, seaming and riveting square or rectangular work. These stakes are available in three sizes: 40, 34-1/2, and 30 inches as measured from the end of the stake to the shank. The *round head stake* is used to finish metal objects that have been bumped out on wood blocks or sandbags. It has a half-spherical head.

Most of the stakes already described fit into a device called a universal holder, figure 18-3. Specially made stakes are used in shops that produce quantity items such as covers. Pieces of iron, figure 18-4, are sometimes used as stakes, depending on the nature of the job.

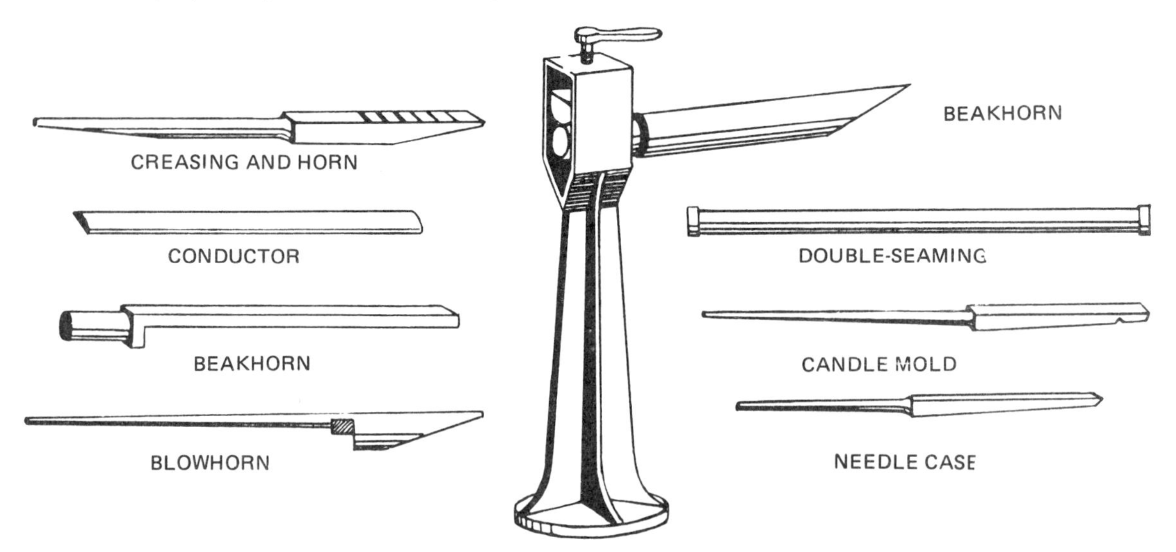

Figure 18-3 Universal holder with stakes.

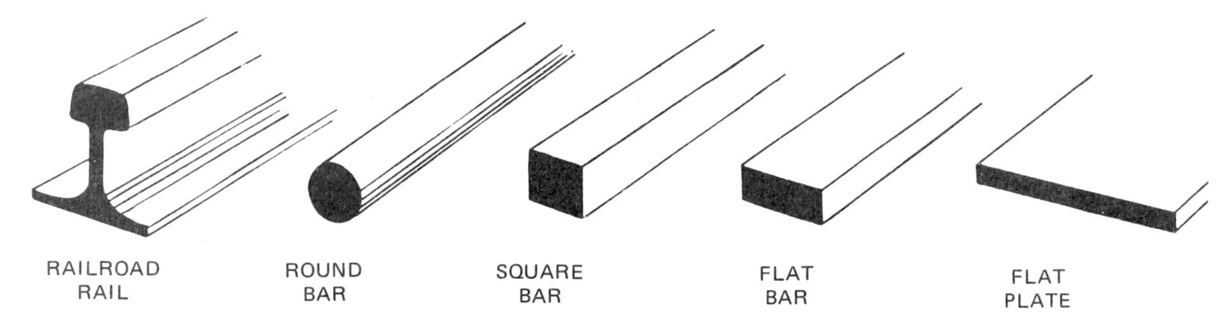

Figure 18-4 Common iron stakes.

THE HAND DOLLY

The hand dolly, figure 18-5, is a portable anvil with a handle which is used for backing or bucking up rivet heads, making double seams, and straightening out kinks.

Figure 18-5 Hand dolly.

Many sheet metal workers carry in their toolbox a hand dolly made of a 2 x 2 inch steel bar. It is about 6 inches long and has one end cut off at a 45-degree angle, about 1/4 inch up from the bottom, figure 18-6.

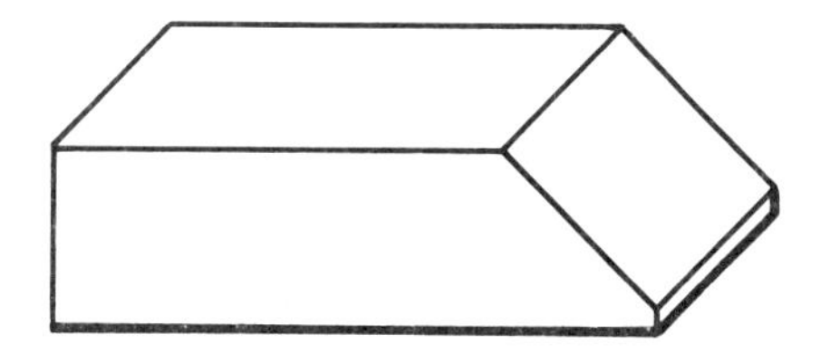

Figure 18-6 Hand dolly.

HOW TO FORM SQUARE JOBS

1. Select the proper stake, such as the beakhorn, needlecase, or hollow mandrel. The stake must be smaller than the inside of the square being formed. Place the stake in the bench plate.

Hand bending small pieces and heavy stock is not recommended because the work is hard to hold.

2. Hold the edge of the metal over the stake with the left hand so that the brake lines are directly over the right-hand edge of the stake, figure 18-7.
3. Holding the metal firmly on the stake with the left hand, bend the extended edge with the right hand to a 90-degree angle.

Figure 18-7
Job on the stake.

Figure 18-8
Job being formed.

4. Hold the metal as shown in figure 18-8 and make a sharp square corner with a mallet. A mallet is used because it covers a large surface and does not stretch the material.

HOW TO FORM ROUND JOBS

1. Select the conductor or needlecase stake and place it in the bench plate. The stake used must be slightly smaller than the diameter of the job.

Figure 18-9
Bending an edge.

Figure 18-10
Bending by hand.

2. Hold the edge of the metal over the stake with the left hand and start the bend with a mallet held in the right hand, figure 18-9. If the job has a grooved seam, insert a piece of scrap material in the groove before bending over the edge with the mallet.
3. Repeat step 2 on the opposite edge.

CAUTION Both starting bends must be in the same direction.

4. Hold either bent edge over the stake with one hand and bend the metal around the stake with the other hand as far as possible, figure 18-10.
5. Reverse the position of the metal and repeat step 4, starting the bend from the other edge.
6. Complete the seam by soldering, grooving, or riveting.
7. Finish shaping the job over the stake by hand or with the mallet.

HOW TO FORM CONICAL JOBS

1. Select a blowhorn or candlemold stake, depending on the size of the job, and place it on the bench plate. For sharp or abrupt tapers, use the apron end of the blowhorn stake. For small, gradual tapers, use the candlemold or blowhorn stake.

2. Hold the edge of the metal over the stake with the left hand, and start the bend with a mallet held in the right hand.

If the job has a grooved seam, be careful not to close the fold for the seam. Insert a piece of scrap metal in the edge before starting the bend.

3. Repeat step 2 on the opposite side.

Both starting bends must be in the same direction.

4. Hold the bent edge over the stake with one hand and bend the metal around it with the other hand, figure 18-11.

Form the work carefully to obtain the proper curve. The shape of the job usually differs from the shape of the blowhorn stake. Therefore, it is good trade practice to follow an imaginary radial line directly over the center line of the stake while bending the metal.

5. Slide the job across the stake until the other edge is over the stake and repeat step 4.

Bend the metal around the stake with a rolling motion. This reduces the possibility of kinks forming in the metal.

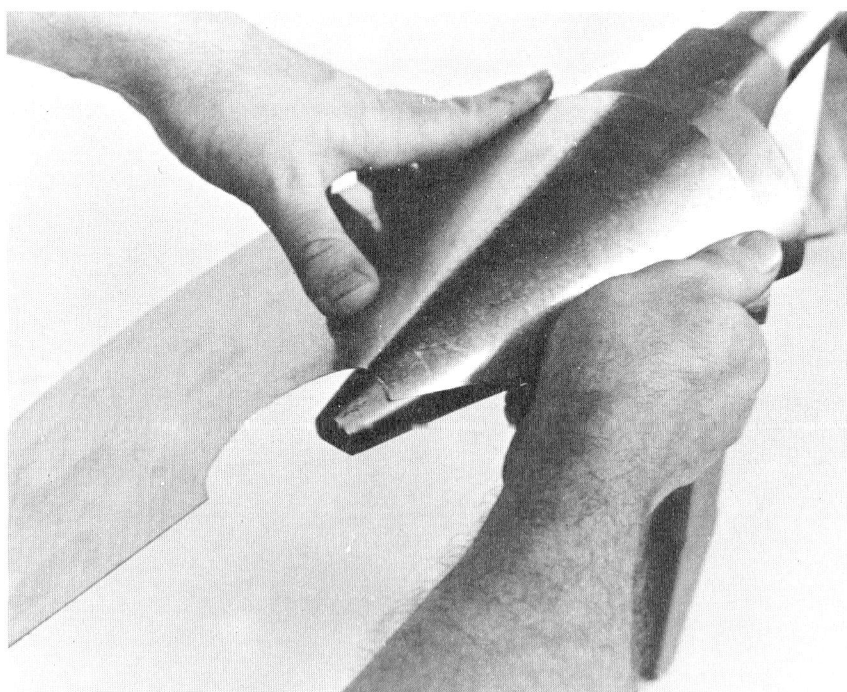

(A) Starting the bend.

Figure 18-11 Forming a conical job.

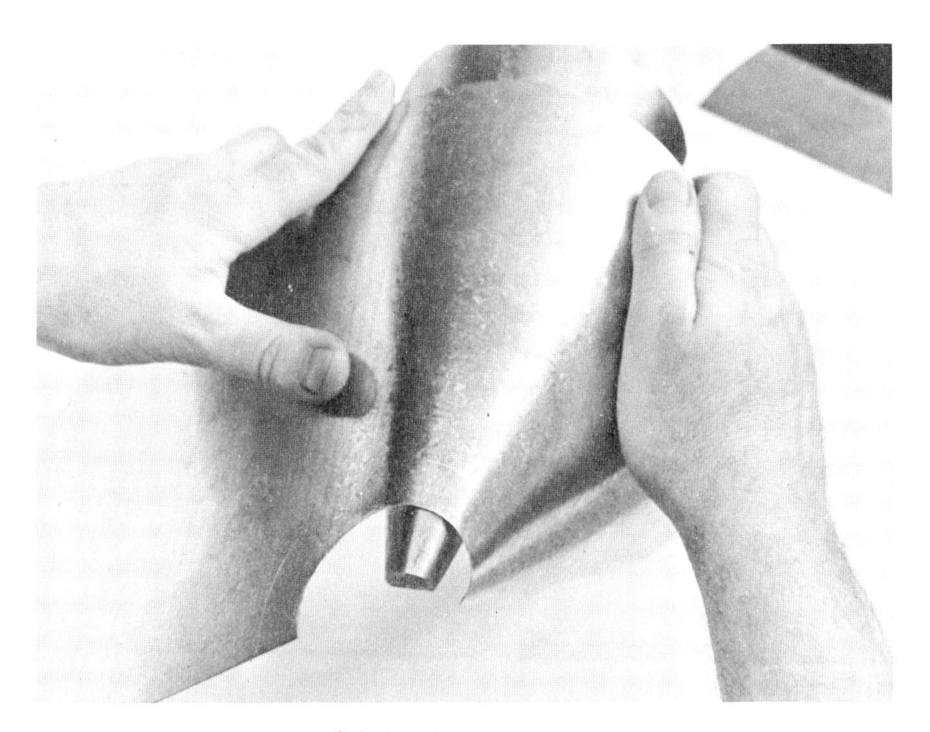

(B) Continuing the bend.

Figure 18-11 Forming a conical job. (Cont'd)

HOW TO FORM WIRE JOBS

1. Select a creasing stake and place it in the bench plate.
2. Select the proper size groove on the stake to fit the wire diameter as closely as possible.
3. If making a pail handle or damper rod requiring a formed end, let the rod extend about 1/4 inch over the edge of the stake and bend it to a 45-degree angle with a hammer, figure 18-12.

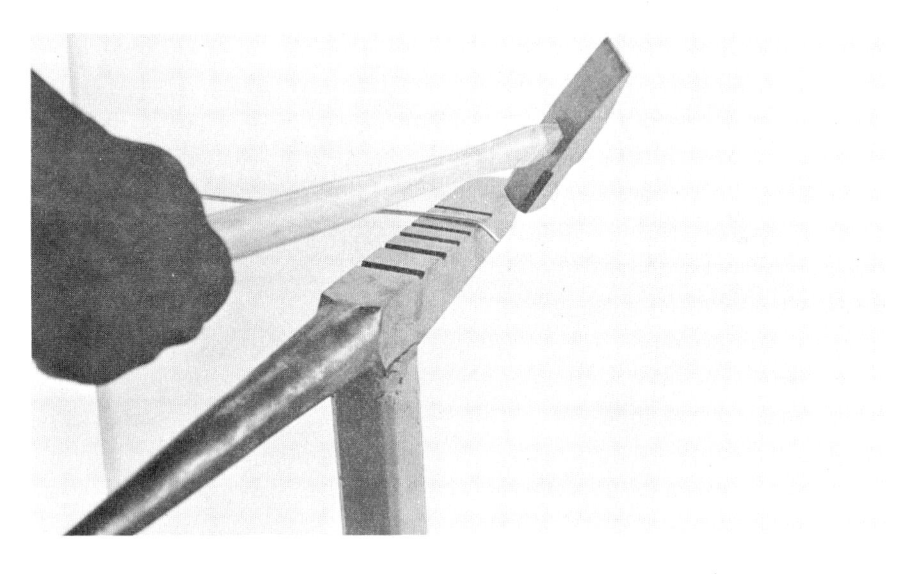

Figure 18-12 The rod is bent at a 45-degree angle approximately 1/4 inch from the end.

Figure 18-13 The rod is bent 90 degrees at the desired length for the loop handle.

Figure 18-14a The rod is bent backwards to finish the loop.

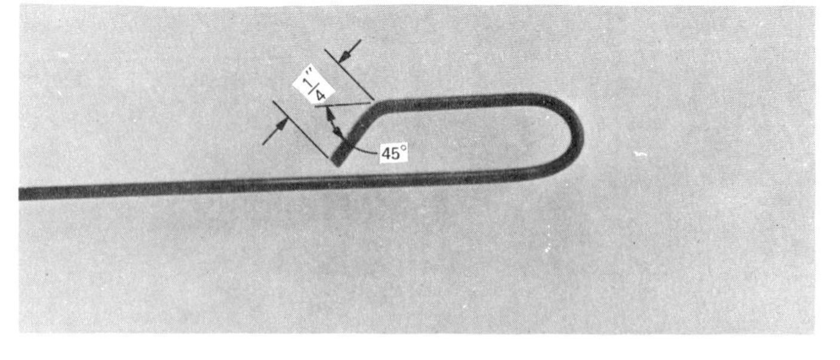

Figure 18-14b The completed damper rod handle.

a. Move the wire forward the desired length and bend it to a 90-degree angle with the hammer, figure 18-13.
b. Place the wire in the groove of the stake so that the 90-degree angle is upright.
c. Strike the wire edge with an inward blow until a loop is formed, figure 18-14.

Figure 18-15a Flat stock for the damper rod holder is grooved on the stake.

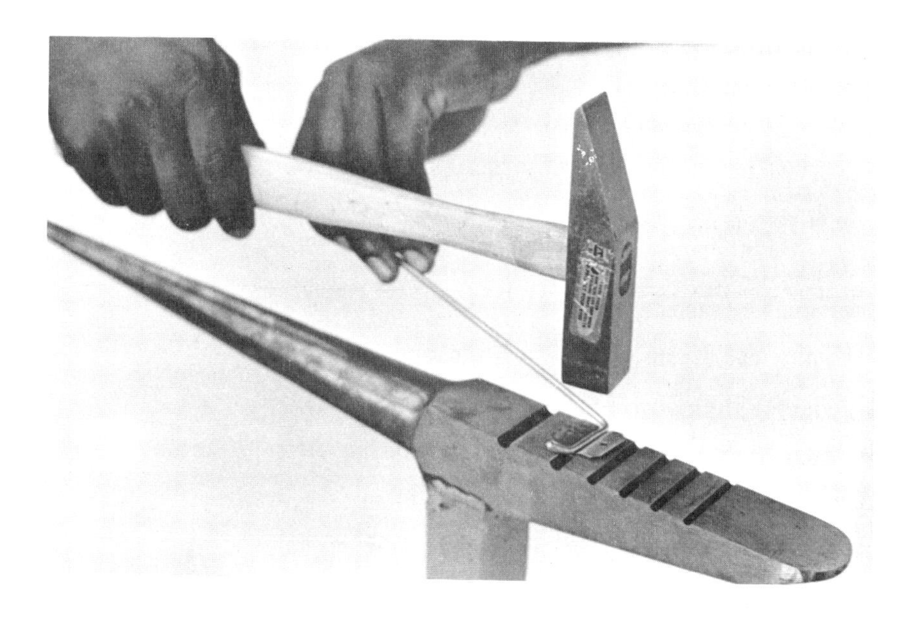

Figure 18-15b The damper rod holder is finished to fit the end of the rod.

4. If a damper rod holder is required, place flat stock over a groove which is slightly larger than the diameter of the damper rod. Form a groove in the stock by placing a piece of wire on top of the stock and hammering both wire and stock into the groove, figure 18-15.

SUMMARY REVIEW

A. Place the answers to the following questions in the column to the right.

1. List the two general uses for a stake. 1. ______________________

____

2. List three shapes that can be formed on a stake. 2. ______________________

____

____

3. What is used to hold a stake in a vertical position? 3. ______________________

____

4. List three different stakes that can be used for riveting. 4. ______________________

____

____

5. A job calls for a collar 4 inches in diameter and 6 inches long. What stake should be selected for forming? 5. ______________________

6. What stakes should be used to make up the cone on a weather cap? 6. ______________________

____

7. A riveted joint is used for installing a large section of pipe in a ventilation system. What device should be selected to back up the rivets when forming the heads? 7. ______________________

8. What can be substituted for a conventional stake if forming a piece of pipe 4 inches in diameter? 8. ______________________

B. Insert the correct word in each of the following sentences.

1. Common ________ stakes are used for chiseling and punching.
2. Many stakes have ________ faces.
3. Bench plates are made of ___________.
4. The __________ stakes are used for general work.
5. When forming, seaming, or riveting pipe, (a, an) _________ stake should be used.

C. Underline the correct word in aech of the following sentences.

1. In order to form the spout on a funnel, a (double-seaming, solid mandrel, blow-horn) stake is used.
2. Wire is formed on a (creasing, beakhorn, hatchet) stake.
3. Forming on a stake should be done with a (peening hammer, rivet hammer, mallet).
4. Hold the piece to be formed on the stake with a (clamp, pair of pliers, hand).

UNIT 19 RIVETING TOOLS

OBJECTIVES

After studying this unit, the student should be able to

- Identify the kinds of rivets used in the sheet metal trade.
- Identify the standard size rivet sets.
- Rivet properly.

One of the most important methods of joining metal together is by riveting. Riveted seams are used where strength is needed and for metals too heavy for seaming machines to form. Hand riveting is done with a hammer and rivet set or with a pneumatic riveting gun and a die or set. Machine riveting is done by a press which squeezes the head on the rivets.

Small jobs are joined by a few rivets; and large jobs, by a series of rivets. In the latter case, the rivets in the ends of the job are upset (expanded by striking the head) to hold the edges of the job in place, and the finish riveting is started at the center hole and continued towards each end. It is necessary to have a solid base when riveting.

RIVETS

Rivets are made of soft, malleable iron which does not crack while the head is being formed. If rivets are coated with tin, corrosion is reduced and the rivets are easier to solder.

There are many shapes of rivets: the tinner's, the flat head, the round head, and the countersunk head are the most common, figure 19-1. The body of the rivet is called the *shank.* The length and diameter of each rivet is measured as shown in figure 19-1. The head of the tinner's rivet is thinner and has a larger diameter than the head of the flat head rivet.

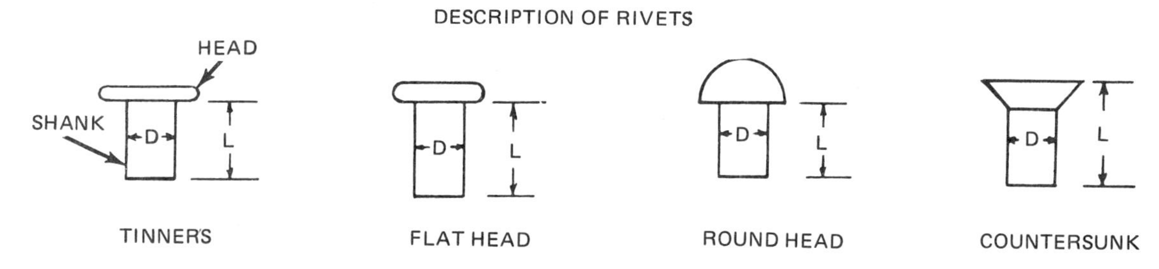

Figure 19-1 Types of rivets.

TINNER'S RIVETS

Tinner's rivets are designated by the weight of 1000 rivets. For example, 8-ounce rivets weigh 8 ounces per 1000, 1-pound rivets weigh 1 pound per 1000, and 2-pound rivets weigh 2 pounds per 1000. As the rivets increase in weight, the diameter and length become greater. Table 19-1 gives sizes of tinner's rivets varying from 8 ounces to 16 pounds. The sizes of coated and uncoated tinner's rivets are the same.

The tinner's rivet selected must be the proper length for a good job. If the rivet is too long, the head cannot be formed correctly and the metal around the rivet is distorted. If the rivet is too short, a complete head cannot be formed.

FLAT HEAD RIVETS

Flat head rivets vary in diameter from 3/32 inch to 7/16 inch in 1/32-inch increments. Rivets of these diameters can be obtained in different lengths. Flat head rivets are needed when more than two thicknesses of sheet metal or pieces of heavier metal are to be riveted together.

Size	Diameter Inches	Length Inches	Size	Diameter Inches	Length Inches
8 oz.	.089	5/32	5 lb.	.186	3/8
10 oz.	.095	11/64	6 lb.	.203	25/64
12 oz.	.105	3/16	7 lb.	.220	13/32
14 oz.	.109	3/16	8 lb.	.224	7/16
1 lb.	.112	13/64	9 lb.	.238	29/64
1¼ lb.	.120	7/32	10 lb.	.238	15/32
1½ lb.	.130	15/64	12 lb.	.259	1/2
1¾ lb.	.134	1/4	14 lb.	.284	33/64
2 lb.	.144	17/64	16 lb.	.300	17/32
2½ lb.	.148	9/32			
3 lb.	.160	5/16			
3½ lb.	.165	21/64			
4 lb.	.176	11/32			

TABLE 19-1 TINNER'S RIVETS

SELECTION OF RIVETS

The rivet selected must be long enough so that the head can be shaped properly. The rivet should project from the job at a distance which is approximately 1 1/2 times the diameter of the rivet.

Example: Find the length of a flat head rivet with a diameter of 5/32 inch which is needed to rivet together a 26-gage sheet and band iron 1/8 inch thick.

Solution:

```
5/32 inch = .1562          .2343
              x 1.5        .0179   (26-gage metal)
              -----       +.1250   (1/8-inch band iron)
               7810       ------
              1562         .3772 =  3/8-inch rivet length, Ans.
             ------
             .23430
```

Table 19-2 gives the sizes of tinner's rivets required to rivet two sheets of metal of the same gage.

Gage of Metal	Size of Rivet	Gage of Metal	Size of Rivet
30 28 26	10 oz. 14 oz. 1 lb.	24 22 20	2 lb. 2½ lb. 3 lb.

TABLE 19-2 TINNER'S RIVET SIZES

RIVET SPACING

The spacing of rivets is given on the drawing or in the specifications of the job. Rivets should not be placed too close to the edge of the seam because they are apt to tear out. The rivet center line should not be less than 2 times the diameter of the rivet from the edge, figure 19-2.

Example: Find the distance that a 1-pound rivet should be placed from the edge of the seam.

Solution: .112 (diameter of rivet: see table 19-1)
x 2
.224 = 15/64 inch from edge, *Ans.*

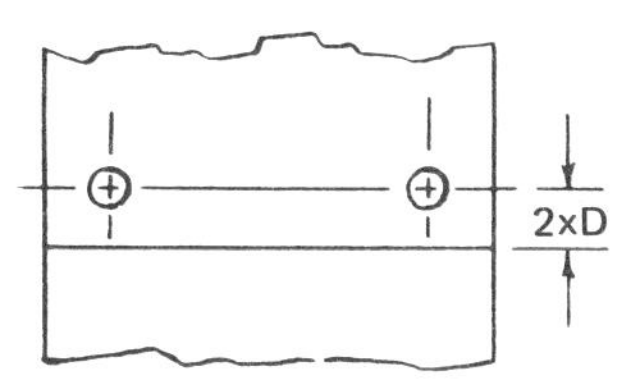

Figure 19-2 Edge allowance.

RIVET SETS

A rivet set is made of tool steel 4 to 6 inches long. The large end has a deep hole and a shallow, cup-shaped hole, figure 19-3. The deep hole fits over the rivet and is used to draw the sheets and the rivet together. The cup-shaped hole is used to form the head on the rivet. A rivet set can be used to force rivets directly through thin metal without previously punching a hole. An outlet at the end of the drawing hole allows the burrs to drop out.

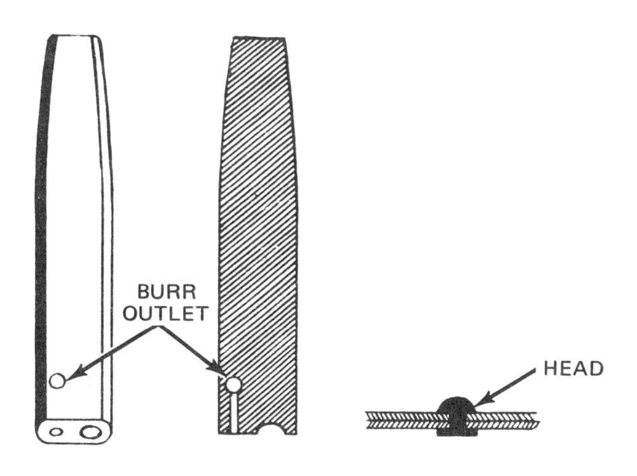

Figure 19-3 Rivet set.

No.	Size Hole In.	For Rivets lbs.	No. (New)	Size Hole In.	For Rivets lbs.
00	5/16	14,16	680	5/16	14
0	9/32	10,12	681	9/32	10,12
1	15/64	7,8	682	1/4	8
2	.2130	6	683	7/32	6
3	.1910	4,5	684	3/16	4,5
4	.1660	3,3½	685	11/64	2½,3
5	.1495	2,2½	686	5/32	1¾,2
6	.1405	1½,1¾	687	9/64	1½
7	.1285	1,1¼	688	1/8	1¼
8	.1100	10, 12 oz.	689	7/64	10,12 oz.

TABLE 19-3 RIVET SETS

Rivet sets are made in a variety of sizes. Table 19-3 gives the sizes of two kinds of rivet sets and the rivets for which the sets are to be used.

SELECTION OF PUNCHES AND RIVET SETS

A rivet set should be selected which is slightly larger than the diameter of the rivet. If it is necessary to punch a hole before riveting, it should also be larger than the diameter of the rivet.

Example: Select tinner's rivets, a punch, and a rivet set for two thicknesses of 26-gage metal.

1. Use a 1-pound rivet for 26-gage metal: see table 19-2.
2. A 1-pound rivet has a diameter of .112 inch: see table 19-1. Use a 1/8-inch solid punch, or a 1/8-inch hand punch.
3. Use a No. 7 or a No. 688 rivet set: see table 19-3.

HOW TO DRIVE RIVETS

1. Punch or drill the rivet hole locations in the work.
2. Place the job to be riveted on a suitable plate or bench stake.
3. Select the rivet and the rivet set.

If the rivet set has a mushroom head, grind it before using.

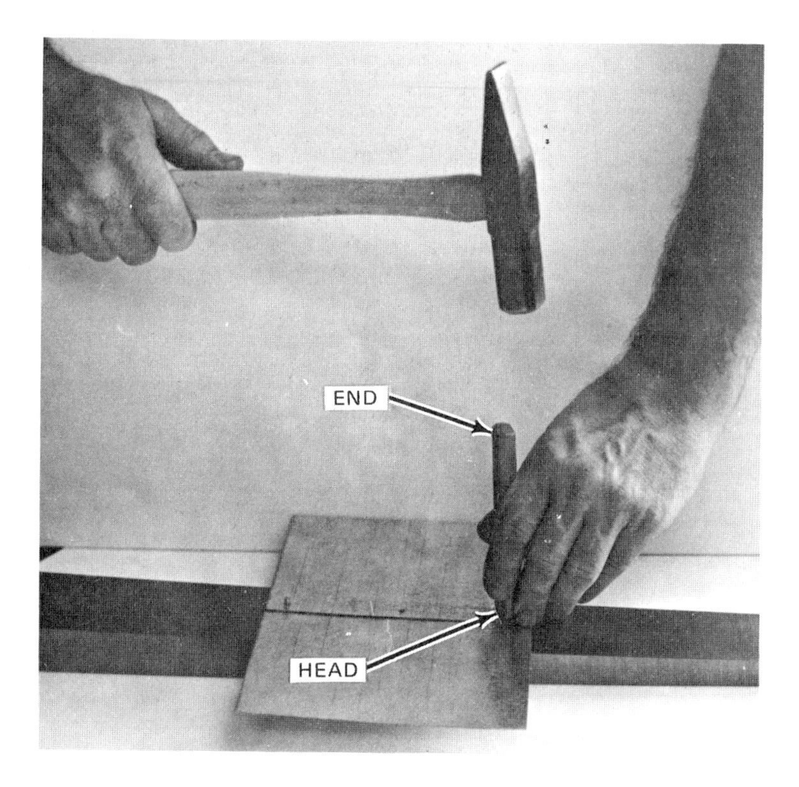

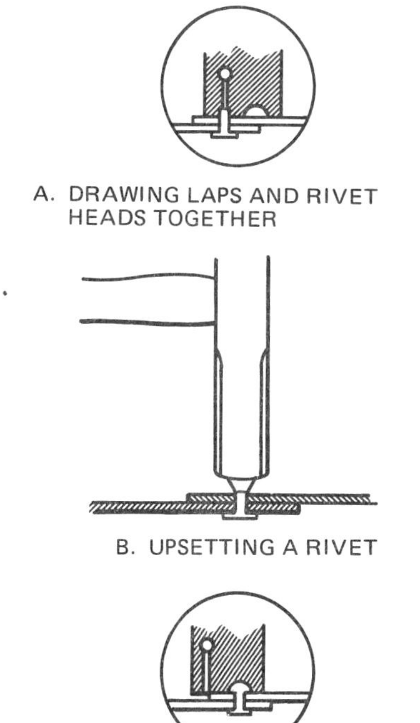

Figure 19-4 Driving a rivet.

4. Select a riveting hammer. Use a light hammer for a small rivet. Make sure the handle is tight.
5. Insert the rivets in the holes, resting the manufactured heads of the rivets on the stake.
6. Draw the laps and the rivet head together tightly by placing the hole in the rivet set over the rivet and striking the set one or two sharp blows with the hammer, figure 19-4A.

Watch the end of the rivet set - not the head of the hammer - to direct the drawing together of the metal and rivet. With practice, one can learn to strike the head of the rivet set without looking at it.

7. Remove the rivet set and upset the rivet by striking the end of the rivet squarely with the face of the hammer, figure 19-4B.

The rivet should not be flattened so much that the metal around the rivet is distorted.

8. To form the driven head of the rivet, place the cup-shaped hole (die) on the rivet, and strike the rivet set with one or two sharp blows with the hammer, figure 19-4C.

Care must be taken to hold the rivet set at right angles to the job to avoid marring the metal. Figure 19-5 shows the results of not drawing the laps and rivet head together. Figure 19-6 shows a rivet properly formed.

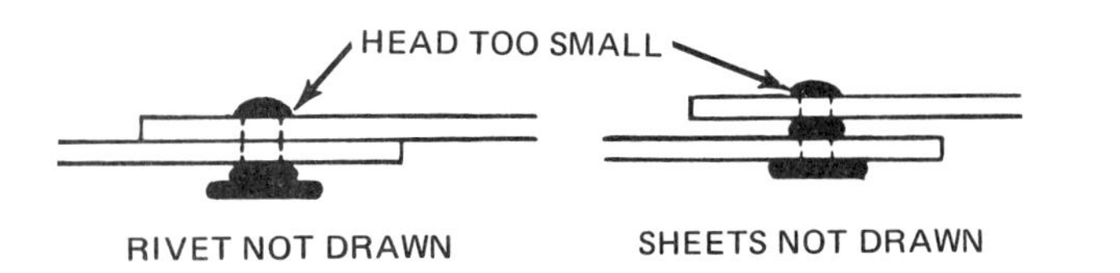

Figure 19-5 Improperly riveted.

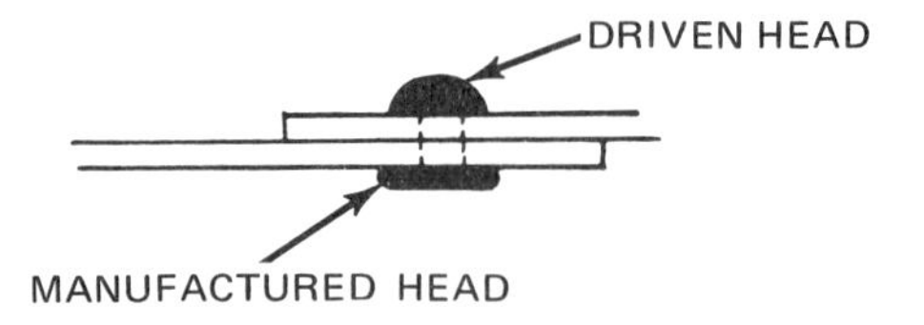

Figure 19-6 Properly riveted.

"POP"® RIVETS

"POP"® rivets, figure 19-7, are used when a leakproof seam is not required. They provide a strong, low cost method of fastening, and are obtainable in steel, aluminum, and Monel®. "POP"® rivets can be used very effectively on light gage metal, but they are not recommended for heavy gage. They are the most useful rivets now made for blind fastenings--where the material being fastened is only accessible on one side. "POP"® rivets are available in sizes of 3/32, 7/64, 1/8, 5/32, 3/16 and 1/4 inch diameters, with countersunk or domed heads.

Figure 19-7 Types of "POP" rivets.

The "POP"® rivet gun shown in figure 19-8 is obtainable with different size nosepieces to fit rivet diameters. These devices usually have two pulling positions and an outside jaw-grip adjustment.

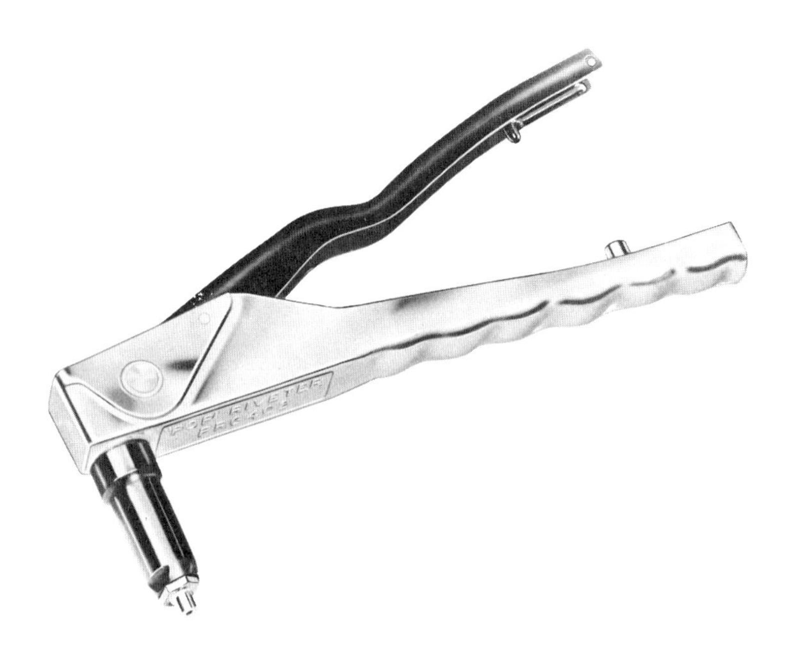

Figure 19-8 "POP" rivet gun.

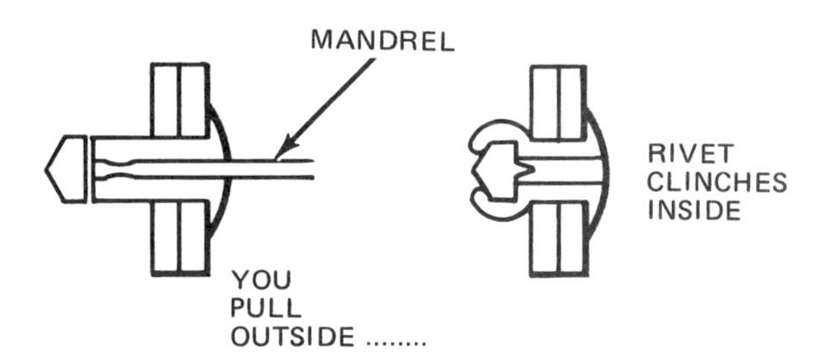

Figure 19-9 Pop rivet action.

HOW TO POP RIVET

1. Select the proper size nosepiece for the rivet to be used and assemble on the gun.
2. Place the rivet mandrel in the nosepiece as far as it will go.
3. Insert the rivet into matching holes of pieces to be joined.
4. Squeeze the handle and pump it several times until the mandrel breaks off.
5. The finished rivet should appear as shown in 19-9.

Rivets can be removed by drilling out their centers with the same size drill as the rivet hole or by cutting off their heads with a cold chisel and hammer (figure 13-12).

SUMMARY REVIEW

A. Place the answers to the following questions in the column to the right.

1. What is the purpose of a rivet? 1. ______________
2. Why must a rivet be the correct length for the job? 2. ______________
3. List the shapes of the more common type rivets. 3. ______________
4. List two reasons for tin plating rivets. 4. ______________
____
5. What does the term 1-pound rivets actually mean? 5. ______________
6. What is a rivet set made of? 6. ______________
7. State the purpose of each hole in the rivet set. 7. ______________
a. the deep hole
b. the cup-shaped hole
____
8. What size rivet set is used for 1-pound rivets? 8. ______________
9. What length rivet is used to join two pieces of 24-gage black iron (.0239 inch)? 9. ______________
10. What would be the edge distance allowance for the rivet in question 9? 10. ______________
11. Briefly list four steps that should be followed in riveting. 11. ______________
____
____
____

B. Insert the correct word in each of the following sentences.

1. Rivets are made of soft, ________________ .
2. Rivets coated with _________ are easier to solder.
3. The edge spacing of rivets should equal _______________________ .
4. (A, An) ______________ rivet should be used to rivet two pieces of 28-gage galvanized iron together.
5. The hole is punched slightly ______________, and a rivet set is selected which is slightly _________ than the rivet diameter.

C. Underline the correct word in each of the following sentences.

1. "POP"® rivets can be used successfully in riveting (heavy gage, light gage) metal.
2. "POP"® rivets can provide a strong fastening in (copper, lead, steel).
3. The "POP"® rivet gun has (one, two, three) pulling positions.
4. The most effective places to use a "POP"® rivet is in (open areas, blind spots, overhead positions).
5. A "POP"® rivet can be easily removed by (punching, pulling, drilling) a hole in its center.

UNIT 20 GROOVING TOOLS

OBJECTIVES

After studying this unit, the student should be able to

- Identify and describe the types of grooving tools.
- Select the proper size hand groover.
- Use the hand groover and grooving rail properly.

The sheet metal worker seldom uses a hand grooving tool because most seaming is done by machine. Nevertheless he should have a knowledge of these tools so that he can use them properly when necessary.

HAND GROOVER

The hand groover is used to offset an outside grooved seam, figure 20-1. This tool is a forging of hardened tool steel with one end recessed to offset the grooved lock. The working surfaces of the groover are accurately finished.

Figure 20-1 Hand groover.
(Courtesy of Pexto.)

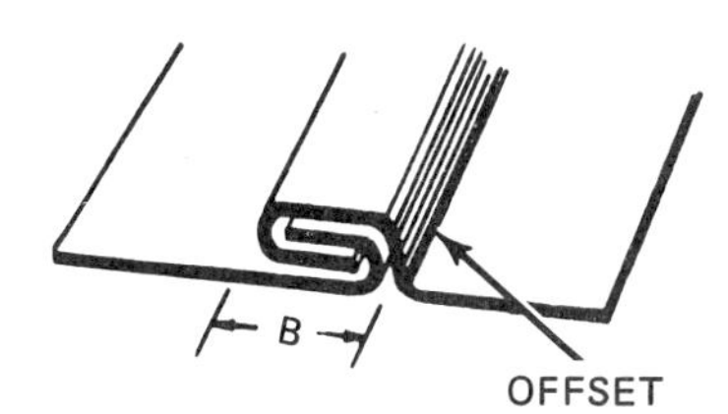

Figure 20-2 Groove seam width.

SIZES

Grooving tools come with grooves ranging from 3/32 to 19/32 inch wide. The groover selected should have a groove about 1/16 inch wider than the width of the seam, figure 20-2. Tables 20-1 and 20-2 give the sizes of two separate sets of groovers in common use.

Number	0000	000	00	0	1	2	3	4	5	6	7	8
Width of Groove In Inches	$\frac{19}{32}$	$\frac{1}{2}$	$\frac{7}{16}$	$\frac{3}{8}$	$\frac{11}{32}$	$\frac{5}{16}$	$\frac{9}{32}$	$\frac{7}{32}$	$\frac{5}{32}$	$\frac{1}{8}$	$\frac{7}{64}$	$\frac{3}{32}$

TABLE 20-1 SIZES OF HAND GROOVERS.

Number	00	0	1	2	3	4	5	6	7	8
Width of Groove In Inches	$\frac{1}{2}$	$\frac{7}{16}$	$\frac{3}{8}$	$\frac{5}{16}$	$\frac{1}{4}$	$\frac{7}{32}$	$\frac{3}{16}$	$\frac{5}{32}$	$\frac{9}{64}$	$\frac{1}{8}$

TABLE 20-2 SIZES OF HAND GROOVERS.

Figure 20-3 Seam properly placed on stake.

HOW TO USE A HAND GROOVER

1. Place the seam over the stake, figure 20-3.
2. Set the seam with a mallet along its entire length.
3. Place the hand groover over the seam in a vertical position.
4. Offset the seam by hammering the hand groover with a rivet hammer or setting hammer at each end.
5. Move along the entire seam length with the groover while hammering, figure 20-4.
6. Dress the seam with a mallet.

Figure 20-4 Seaming with a hand groover.

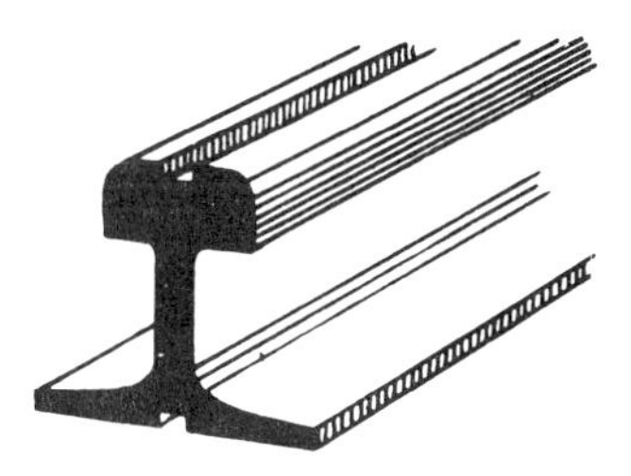

Figure 20-5 Grooving rail.

GROOVING RAIL

A grooving rail is a piece of railroad rail with a groove milled in the top and the bottom faces, figure 20-5. The top is used for round work; and the bottom, for flat or square work. It is used to offset an inside grooved seam; the bulge of the seam fits in the groove in the rail. Several rails with different widths of grooves should be available in every shop. The widths of the grooves should be the same as the widths of corresponding hand groovers.

HOW TO USE THE GROOVING RAIL

1. Select the proper size rail for the required seam width.
2. Fasten the rail to the bench with clamps.
3. Place the seam, hooked together, over the rail and in the grooves. Hold it tightly.
4. Strike one end and then the other end with a mallet.
5. Strike along the entire length of the seam with a mallet.
6. Remove the seam from the groove. Place it on a flat surface and close it by striking with a mallet.

SUMMARY REVIEW

A. Place the answers to the following questions in the column to the right.

1. What is the purpose of the hand groover? 1. ____________
2. What material is used in manufacturing a hand groover? 2. ____________
3. How is a hand groover selected for a seam width? 3. ____________
4. What is the range of sizes for the common hand groover? 4. ____________
5. What is the purpose of a grooving rail? 5. ____________
6. How do the grooves in a rail compare with those of a hand groover? 6. ____________

B. Insert the correct word in the following:

1. A grooved seam is offset at ______ ends before working the entire length.
2. A mallet is used to dress the seam in order to prevent ____________.

C. Underline the correct word in each of the following sentences.

1. A grooving rail is used to finish an (outside, inside) grooved seam.
2. The hand groover is struck with a (mallet, rivet hammer, stake) to set the offset along the seam.

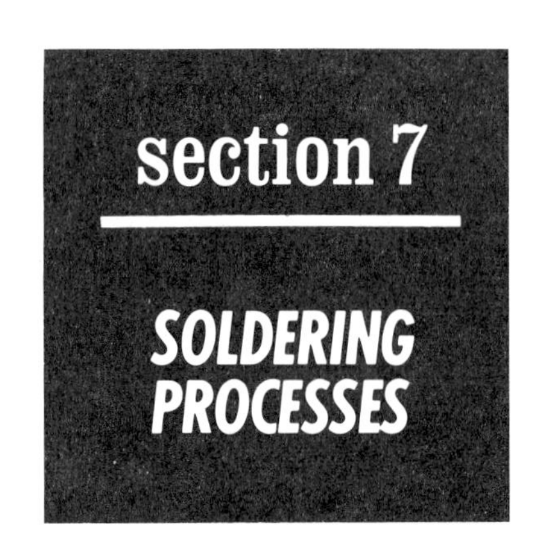

UNIT 21 *SOLDER FURNACES*

OBJECTIVES

After studying this unit, the student should be able to

- Identify and describe the types of soldering furnaces used in the shop and in the field.
- Light each type of furnace safely.
- Select the best type of furnace for the job at hand.

Portable and stationary heating equipment is used in the sheet metal trade for heating fire-heated soldering irons. The common types are the bench gas furnace in the shop, and the gasoline fire-pot, propane tinner's furnace, and charcoal pot in the field. The types of fuel used in solder furnaces and precautions for their use are given in table 21-1.

FUEL	STORAGE METHODS	SAFETY PRECAUTIONS
Charcoal	Bins or bags	Keep away from open flame.
Propane	Metal tanks	Keep away from heat and flame. Do not throw empty tanks in fire. Check all valves and connections for leaks.
Gas	Manufacturer's responsibility	Check all valves and connections for leaks. Avoid accumulation in furnace.

TABLE 21-1 SOLDER FURNACE FUELS

GASOLINE FIRE POT

The gasoline fire-pot uses an air pressure system to force fuel into a chamber which is kept just hot enough to vaporize the fuel.

HOW TO LIGHT THE GASOLINE FIRE POT

1. Stroke the air pump five or six times.
2. Open the control valve slightly and let the cup fill.

Do not overfill the cup or spill gas. Wipe up any spills and allow to dry before igniting the pot.

3. Ignite the gasoline cup with a match.
4. When the flame begins to die down, open the valve slowly. The burner will ignite.
5. Regulate the control valve to the desired flame.

PROPANE TINNER'S FURNACE

As the name states, this furnace, figure 21-1, uses propane gas as the fuel. The fuel tanks have an excess pressure relief valve and a flow check valve that provide safety in its operation.

Figure 21-1 Propane tinner's furnace.

Figure 21-2 Tinner's charcoal pot.

CHARCOAL POT

The tinner's charcoal pot, figure 21-2, is one of the oldest and still most serviceable portable heaters in use. Heat can be controlled in all kinds of weather by raising or lowering the lid or adjusting the damper to regulate the air flow through the charcoal.

HOW TO LIGHT THE CHARCOAL POT

1. Crumble paper and place it in the bottom of the pot.
2. Fill the pot about one-third full with medium size charcoal.
3. Light the paper with a match.
4. When charcoal begins to burn freely, fill the pot.
5. Partly close the cover or damper.

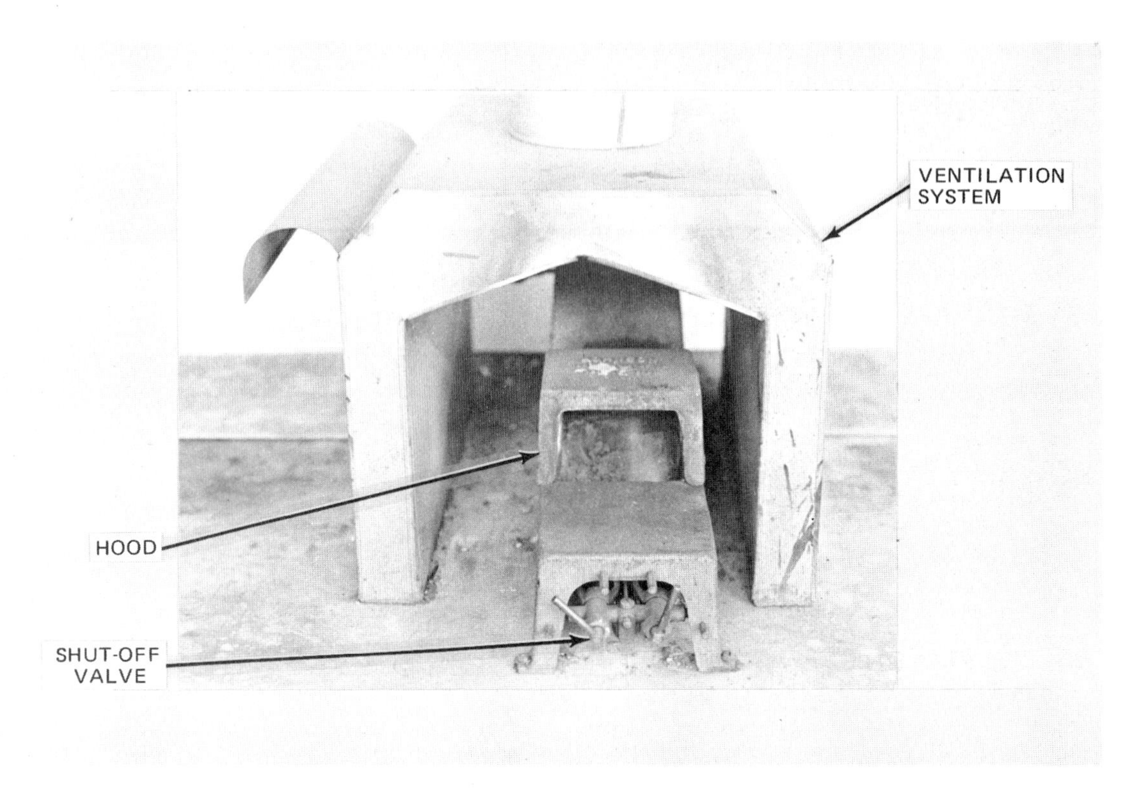

Figure 21-3 Gas furnace.

GAS FURNACE

The best method of heating soldering coppers in the shop is with a gas furnace, figure 21-3. This furnace consists of a gas stove with a hood. The hood can be raised so that the bed of the furnace can be cleaned easily.

A refractory lining inside the furnace prevents the metal part of the furnace from being burned. The hood and the lining is shaped to deflect the heat over the soldering coppers.

Most furnaces are equipped with two burners controlled by separate valves which can be adjusted to the desired temperature. The furnace also has a pilot light which is allowed to remain burning during the working period when the furnace is not in use.

The furnace will quickly produce a heat of 1800 degrees Fahrenheit. This is sufficient to heat 10-pound soldering coppers.

HOW TO OPERATE THE GAS FURNACE

1. Check to see that the main line valve, the branch line valves, and the furnace shutoff valves are turned off.
2. Remove the dirt from inside the gas furnace.
3. Turn on the main line valve.
4. Light a small piece of twisted paper and place it in the mouth of the furnace.
5. Turn on the branch line valve, thereby lighting the pilot light on the furnace.
6. Open one of the shutoff valves located at the front of the furnace by turning the valve to a vertical position, figure 21-4.

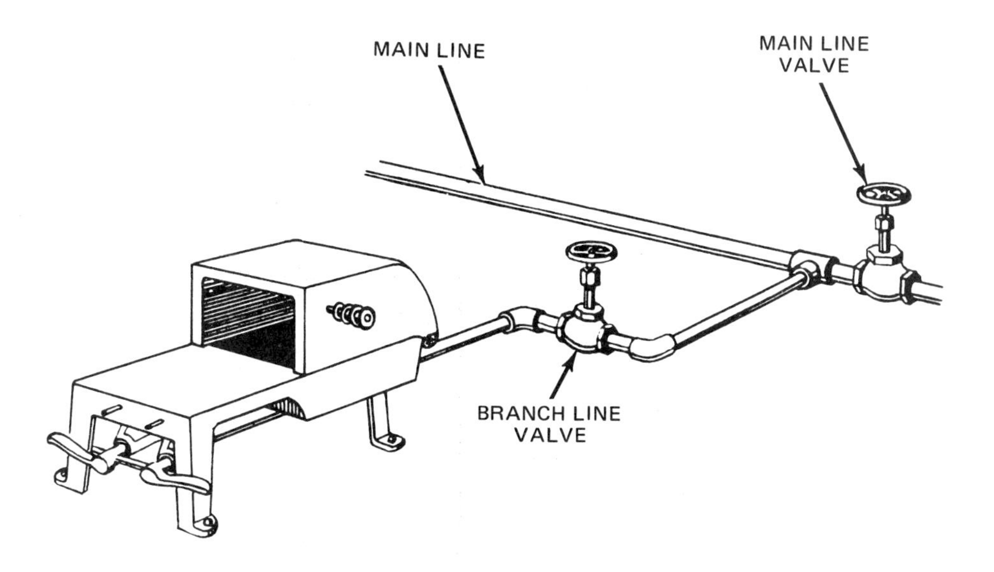

Figure 21-4 Furnace and valves.

7. Turn on the other shutoff valve at the front of the furnace.
8. Adjust the flame to maintain the heat needed by turning the shutoff valves. A full flame may be used, and the soldering coppers may be positioned as necessary to be heated to the desired temperature.
9. When not in use, the furnace shutoff valves must be turned off, leaving only the pilot light burning.
10. At the end of each day, turn off the furnace shutoff valves, the branch line valves, and the main line valve.

Make certain that the valves are turned off. An accumulation of gas is very dangerous and may cause a serious explosion.

SUMMARY REVIEW

A. Place the answer to the following questions in the column to the right.

1. List three portable furnaces for heating soldering coppers. 1. __________
   ____
   ____
2. What type of furnace should be selected for heating soldering coppers on a roof flashing job? 2. __________
3. On a windy day what portable heater is most practical for heating soldering coppers? 3. __________
4. List the five principal parts of a gas soldering copper furnace. 4. __________
   ____
   ____
   ____
   ____

5. What are two purposes of the refractory lining inside the gas furnace? 5. ____________________

____

B. Insert the correct word in each of the following sentences.

1. The gasoline fire pot uses (a, an) ____________ system to force fuel into the chamber.
2. Adequate ____________ is important to prevent fire when using the gasoline fire pot.
3. The ____________ has the advantage of heat control in all kinds of weather.

C. Underline the correct word in each of the following sentences.

1. The gas soldering furnace can quickly produce a heat of (2500 degrees Fahrenheit, 1800 degrees Fahrenheit, 2800 degrees Fahrenheit).
2. At the end of a working day, make sure that gas furnace valves are shut to prevent a gas accumulation that may cause (injury, damage, explosion).

UNIT 22 SOLDER AND FLUXES

OBJECTIVES

After studying this unit, the student should be able to

- List the classes of solder and types of fluxes.
- Identify the elements of composition in solder.
- Select the proper solder and flux for the job.

Solders may be divided into three classes: soft solders, hard solders, and aluminum solders. Soft solders and occasionally aluminum solders are applied by means of a soldering copper; hard solders and aluminum solders are applied with a blowtorch. This unit is concerned with soft soldering, which is the method used most extensively by the sheet metal worker.

Soft solder is an alloy composed of tin and lead. Sometimes other metals are added to the alloy to increase its strength or change its melting point.

SOFT SOLDERS

Soft solders are low in strength and have comparatively low melting points. They are easily applied and flow readily because of the low melting points. The melting point of the alloy is less than that of each metal itself. For instance, tin melts at 456 degrees Fahrenheit and lead melts at 621 degrees Fahrenheit, but a 50-percent mixture of each melts completely at about 415 degrees Fahrenheit. When referring to percentages of tin and lead, tin is always expressed first.

The sheet metal trade uses a 50-percent tin and 50-percent lead solder (called half and half). This solder is available in bars of different shapes from 1/4 to 1 1/2 pounds.

When the proportion of lead and tin is changed, the melting point of the alloy is changed, table 22-1. A solder, made of 60 percent tin and 40 percent lead with a melting point of about 370 degrees Fahrenheit, is used for soldering light gage metal. Since the melting point is lower than that of the half-and-half solder, the temperature of the copper and the work can be lower and there is less chance of the work buckling. If the proportion of lead is increased, the melting point of the alloy becomes higher (up to 621 degrees Fahrenheit) and the solder does not flow as readily.

Composition (percent)	Melting Point	Composition (percent)	Melting Point
10/90	573	60/40	374
20/80	533	70/30	376
30/70	496	80/20	396
40/60	460	90/10	421
50/50	418		

TABLE 22-1 TIN-LEAD MELTING POINT

Most of the solders in the form of wire are composed of 60-percent tin and 40-percent lead. They come in spools and may be solid solder or have an acid or rosin core. The most popular size is about 1/8 inch in diameter.

Soft solders are easily applied and flow readily, but the work to be soldered must be cleaned, the proper flux and soldering copper used, and the copper properly tinned and heated to the correct temperature.

Pure tin (also called block tin) is used when soldering kitchen utensils and food processing equipment because of the danger of lead poisoning. It has a brighter appearance than solders composed of lead and tin.

FLUXES

The strength of a soldered joint depends on the adherence of the solder to the metal being joined. To obtain good adherence, it is necessary that the surface of the metal and the solder be free of oxide and dirt. Metals are usually covered with a film of oxide, and the amount of oxide increases as the metal is heated to the soldering temperature. Flux removes the oxide film already present and protects the surfaces of both metal and solder from the air while they are heated to the soldering temperature.

Various fluxes are used when soldering different kinds of metal, table 22-2. The fluxes ordinarily used for soft soldering are solutions or pastes that contain *zinc chloride.* Zinc chloride is commonly called *boiled acid, cut acid,* or *killed acid.* It is odorless and colorless. It should be labeled on its container for safety.

FLUX	METAL SOLDERED
Muriatic Acid (raw acid)	Galvanized Iron Dull Brass Dull Copper
Zinc Chloride (cut acid)	Black Iron Copper Brass Iron Zinc Monel Tarnished Tin Plate
Rosin	Tin Plate Lead Bright Copper Terne Plate
Phosphoric Acid	Stainless Steel

TABLE 22-2 SOLDER FLUXES

The substance holding the flux material is evaporated by the heat of the soldering operation, leaving a layer of solid flux on the work. At the soldering temperature, the solid flux is melted and partially decomposed so that it liberates hydrochloric acid. This acid then dissolves the oxides from the surfaces of the solder and the work. The melted flux also forms a protective film on the work that prevents further oxidation from taking place.

Zinc chloride cleans and prevents oxidation when soldering black iron. Because zinc chloride fluxes have a corrosive action, it is sometimes necessary to employ a noncorrosive flux for work when the last traces of the flux cannot be removed after the soldering is completed. *Rosin* is the most commonly used flux of this type.

Muriatic acid, also called *raw acid,* is the commercial form of hydrochloric acid and is yellow in color. This acid is used as a flux when soldering galvanized iron, but it is good trade practice to add a little zinc to the raw acid to prevent blackening the galvanized iron around the soldered joint. Muriatic acid is not used on black iron because it dries and curls up when touched with a heated soldering copper.

Rosin is used as a flux on new tin plate and terne plate because it is noncorrosive. Rosin only prevents oxidation and does not clean tin plate. Therefore, old tin plate is cleaned with muriatic acid, but rosin is used for actual soldering.

Prepared soldering salts, which must be dissolved in water before being used, are also available. These salts do not form acid when dissolved and can be used to solder any of the above metals. At present there are many good commercially prepared fluxes. Some are stainless, and others are nonacid. The manufacturer's recommendations should be followed in all cases.

HOW TO PREPARE ZINC CHLORIDE

1. Put muriatic acid into an earthenware jar.

Place the jar outdoors or near an open window.

2. Add small pieces of zinc, a few at a time, to the acid. This will avoid excessive boiling.
3. Continue adding zinc until all boiling stops and a few pieces of zinc are left in the jar.
4. Skim or filter the solution through a cloth.

The fumes are injurious when inhaled and are inflammable. If acid comes in contact with the skin or eyes, wash immediately with cold water. Consult a doctor.

DIPPING SOLUTION

A dipping solution is used to clean soldering coppers. During use, a copper becomes discolored by oxidation. By dipping it in the solution, the surface becomes bright and clean. The copper can then hold the solder and allow it to flow evenly to metals.

HOW TO PREPARE A DIPPING SOLUTION

1. Put 1 pint of water in a small bowl, earthenware crock, or glass container.
2. Add 1/8 ounce powdered sal ammoniac.
3. Stir until the powder is dissolved.

A weak solution made by dissolving salts in water may be used as a dipping solution. However, the acid or flux that is used for the soldering process should never be used as a dipping solution.

HOW TO APPLY A FLUX

Liquid

1. Obtain a small brush or swab and dip it into liquid flux.
2. Spread flux lightly only on the place to be soldered.

Carelessness in applying flux can cause serious burns and eye injury. Do not flip or drop the brush.

Powder

1. Sprinkle powdered flux on the place to be soldered or melt it on the metal with a hot soldering copper.

SUMMARY REVIEW

A. Place the answers to the following questions in the column to the right.

1. List the three classes of solder now in use. 1. ______________________
   ____
   ____
2. List the two elements that are alloyed to make soft solder. 2. ______________________
   ____
3. How does the addition of a third element change the characteristics of a basic soft solder alloy? 3. ______________________
4. List the advantages that soft solder has for the sheet metal worker. 4. ______________________
   ____
   ____
5. Describe the composition of the two most widely used soft solders. 5. ______________________
   ____
   ____
   ____
6. What is the melting point of each of the soft solders in question 5? 6. ______________________
   ____
7. What soft solder is most suitable for soldering light gage copper? 7. ______________________
8. Give two reasons why the solder named in question 7 should be selected. 8. ______________________
   ____
9. Identify two types of wire solder. 9. ______________________
   ____
10. Why is pure tin used as a solder on all kitchen utensils and food processing equipment? 10. ______________________

B. Insert the correct word in each of the following sentences.

1. The basic functions of a soldering flux is to remove _______ and __________ surfaces from the air.
2. Fluxes are obtainable in both the _______ and _________ states.
3. The most commonly used non-corrosive flux is ___________ .
4. Old tin plate should be cleaned with ____________________ before soldering.

C. Underline the correct word in each of the following sentences.

1. Galvanized iron should be soldered using (zinc chloride, rosin, muriatic acid) as a flux.
2. A zinc chloride flux should be prepared in a (tin can, earthenware jar, metal pot).
3. A dipping solution should be composed of water and (muriatic acid, powdered sal ammoniac, rosin).
4. If acid contacts the skin, it should be (wiped, washed) off immediately.

UNIT 23 SOLDERING COPPERS

OBJECTIVES

After studying this unit, the student should be able to

- Identify the commonly used shapes of soldering coppers.
- Describe the conditions which require shaping, cleaning, and finishing a soldering copper.
- Select the proper soldering copper for the job and use it correctly.

The soldering copper transmits heat to a piece of metal in order to make a firm and permanent bond between the metal and the solder. It is often called a soldering *iron.* A soldering copper consists of a forged piece of copper connected by an iron rod to a handle. The copper end is called the *bit* or *head.*

Copper is metal which absorbs and conducts heat readily. It can be easily forged to any shape without changing the properties of copper. Copper tins readily and holds the tinning while being reheated, provided the copper is not overheated. (See unit 24). Both wood and fiber handles are available for soldering coppers. The standard wood handle is forced on the rod, and a special kind of handle, the Shur-Grip®, is screwed on the rod.

TYPES OF SOLDERING BITS

There are four types of soldering copper bits: the roofing copper, the hatchet copper, the pointed copper, and the bottom copper. The pointed copper and the bottom copper are used most often in the sheet metal shop.

Figure 23-1 shows a pointed copper: the end is shaped like a pyramid and has a dull point. The copper head is either forged on the end of an iron rod or attached between the split ends of the rod.

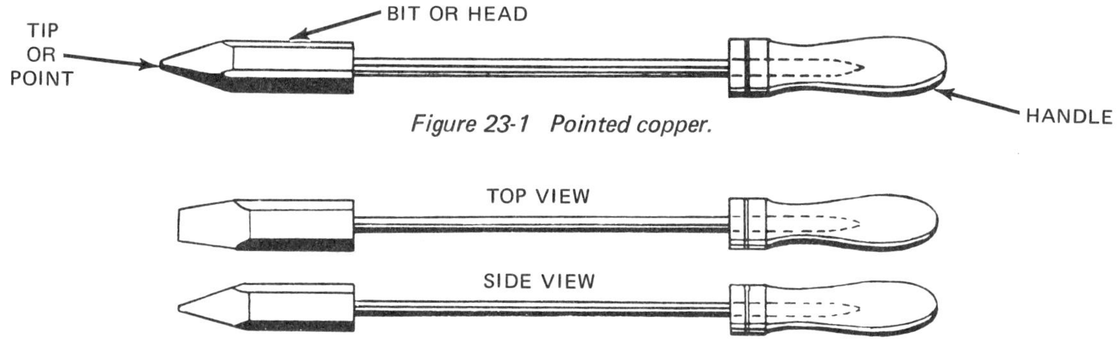

Figure 23-1 Pointed copper.

Figure 23-2 Bottom copper.

The bottom copper has a broad tip shaped like a wedge, figure 23-2. It is used to solder bottoms in tanks and containers of all kinds. It may be used in any place where the copper must be held in an upright position. Ordinarily, soldering is done with the copper in an inclined position, and the metal is preheated by the main part of the tip as the soldering progresses, figure 23-3A. Because the tip of the bottom copper is broad, however, the copper can be used in an upright position, and the work is still preheated by the broad part of the tip, figure 23-3B.

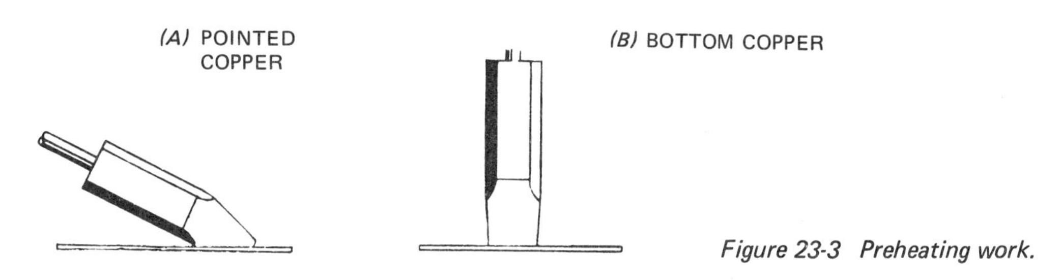

Figure 23-3 Preheating work.

SIZES OF SOLDERING COPPERS

Soldering coppers are classified by the weight in pounds of two copper bits; thus, a pair of 2-pound coppers has two copper bits, each weighing 1 pound. The common sizes of soldering coppers are 1, 1 1/2, 2, 3, 4, 5, 6 and 8 pounds per pair. The 12- and 16-pound coppers are used for roofing jobs and heavy metal operations.

A light copper is never used for heavy gage metal. It does not hold enough heat to heat the metal or to allow the solder to flow properly. Heavy coppers are inappropriate for light gage metal. They are clumsy, and the additional heat causes the metal to buckle.

PREPARATION OF A COPPER BIT

The sheet metal worker must be able to forge and finish a copper bit when necessary. The points of soldering coppers are often reshaped by means of forging or filing to suit a particular job. Soldering coppers must be filed to a smooth finish when any one of the following conditions exist:

- burned surface from overheating
- scale or corrosion
- old tinning
- pit holes

HOW TO SHAPE A SOLDERING COPPER BY FORGING

1. Heat the soldering copper to a bright red.
2. Clamp the copper in the vise and file it until all the burned tinning and pits are removed.

Keep your fingers away from the hot soldering copper.

3. Reheat the soldering copper to a bright red.
4. Holding the copper on an anvil or heavy iron plate with the left hand, forge the copper to the required shape by striking the copper with a large hammer, figure 23-4. The point of the soldering copper is hammered back as the forging progresses. The point should be blunt, and the head should not be tapered too much so that it will hold heat during soldering. Turn the copper often to produce a square surface. If the soldering copper cools off before the required shape is obtained, reheat the copper as often as necessary.

Figure 23-4
Forging a soldering copper.

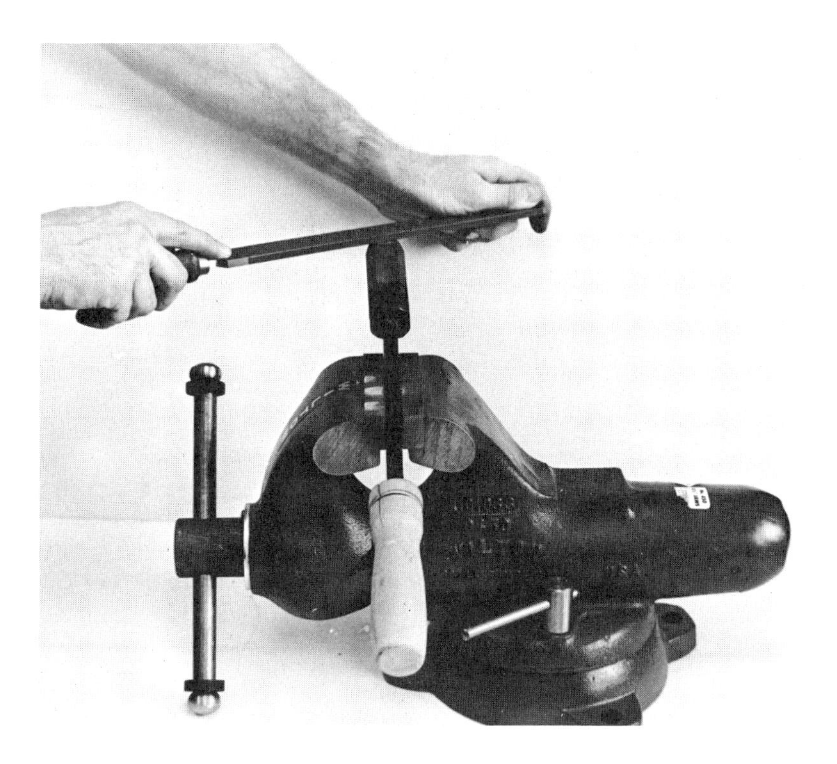

Figure 23-5 Filing a copper.

HOW TO SHAPE, CLEAN, OR FINISH A SOLDERING COPPER BY FILING

1. Check the gas furnace to see that all valves are shut off.
2. Light the gas furnace. (See unit 21, page 132.)
3. Make sure that the handles on the soldering coppers are secure. To renew a standard wooden handle, heat the square pointed rod and push it quickly into the handle. Remove the rod, allow it to cool, and then force the handle on the rod.
4. Place the soldering copper in the gas furnace until it is heated to a cherry red.
5. Remove the soldering copper from the furnace and clamp it in a vise.
6. Grasp a 12-inch, single-cut bastard file as shown in figure 23-5. File the copper by bearing down on the forward stroke and releasing the pressure on the return stroke.

Avoid rocking the file or the surface will be uneven. Keep your fingers away from the hot soldering copper.

7. File the pointed sides of the soldering copper until they are bright and smooth on all four sides.
8. File off the sharp point of the soldering copper.
9. Reposition the copper in the vise and file off the sharp edges as shown in figure 23-6.

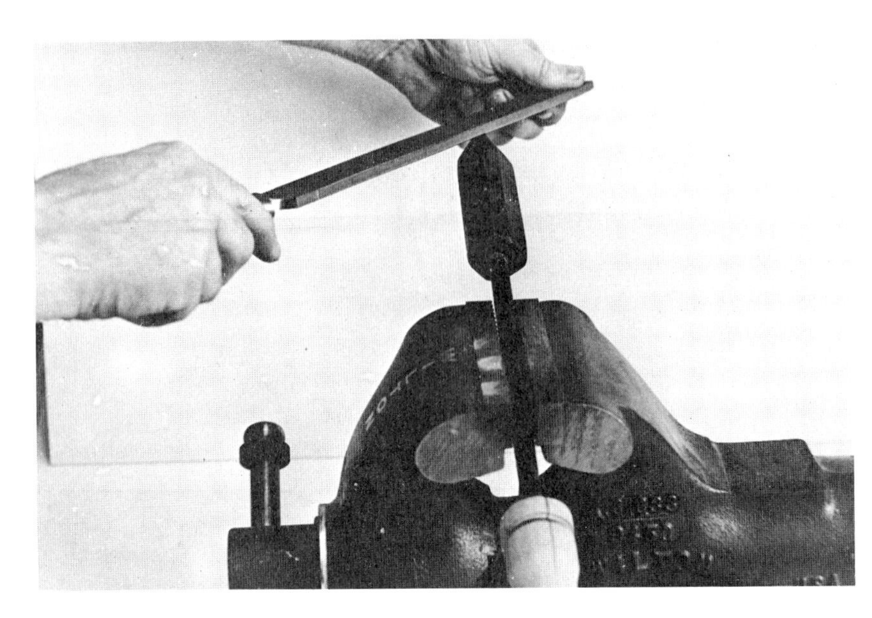

Figure 23-6
Filing corners of copper.

SUMMARY REVIEW

A. Place the answers to the following questions in the column to the right.

1. List two commonly used shapes on soldering coppers. 1. ______ ______
2. List the three parts of a soldering copper. 2. ______ ______ ______
3. What is the main use of a bottom copper? 3. ______
4. Give three reasons why copper is used as the soldering end of soldering coppers. 4. ______ ______ ______
5. List five conditions that would require filing of soldering coppers. 5. ______ ______ ______ ______ ______

B. Insert the correct word in each of the following sentences.

1. If too large a soldering copper is selected for a job, it can cause ______ .
2. A soldering copper should be heated to ______ for forging.
3. If a soldering copper is forged to a long tapered point, it ______ rapidly while soldering.

C. Underline the correct word in each of the following sentences.

1. The bit of a 4-pound soldering copper weighs (2, 4, 8) pounds.
2. The preferred shape of a soldering copper for general use is a (long taper, blunt taper, chisel point).

UNIT 24 *SOLDERING METHODS*

OBJECTIVES

After studying this unit, the student should be able to

- List the three methods of soldering in common use.
- Tin a soldering copper properly.
- Solder sheet metal seams properly.

Many good jobs are spoiled by poor soldering. Careful attention to details and much practice are necessary to produce a satisfactory job. The flux and copper chosen must be suitable for the type of metal and the seam to be soldered. In addition, the copper must be well tinned and heated to the correct temperature.

TERMINOLOGY

Soldering is the joining together of two or more pieces of metal by means of an alloy having a lower melting point than the pieces being joined. Soldering differs from welding in that soldering does not penetrate the base metal.

Sweating is a type of soldering in which the correctly heated copper is held in the proper position on a joint so that the solder flows completely through the joint. In some cases the joint is pretinned, that is, the metal is covered with a thin coat of solder before it is joined by sweating.

Skimming is the pulling of a thin coat of solder over the surface of a joint. It usually leaves a weak joint.

Tacking is the holding of a joint in its correct position by melting small drops of solder at intervals along the seam. The joint is finished by sweating.

Tinning is the application of solder to the tip of a clean, heated copper. Tinning enables the molten solder to adhere to the copper during use.

HOW TO TIN A COPPER

1. File the soldering copper, if necessary, as instructed in unit 23, page 141.
2. Place the copper in the furnace and heat it to a degree that will just melt the solder.
3. Remove the copper from the furnace and rub each side back and forth across a cake of sal ammoniac to clean it. Sal ammoniac is muriate of ammonia and has a chemical cleaning action on hot copper. It is obtainable in 1/4- to 1-pound blocks under such trade names as Salamac® and Speco®.

The iron should be hot enough to cause white fumes to rise from the sal ammoniac block. The fumes are a toxic gas and should not be inhaled.

4. Add a little solder to the soldering copper while rubbing each side of it back and forth on the sal ammoniac until the copper is tinned, figure 24-1.
5. The soldering copper is now ready for use. Place it in the furnace to maintain the proper temperature if there is a delay before soldering.

The soldering copper should not get red hot because excessive heat burns the tinning and oxidizes the copper so that it must be filed, cleaned, and retinned. A brilliant green flame in the furnace at the tip of the copper indicates that the tinning is being burned off.

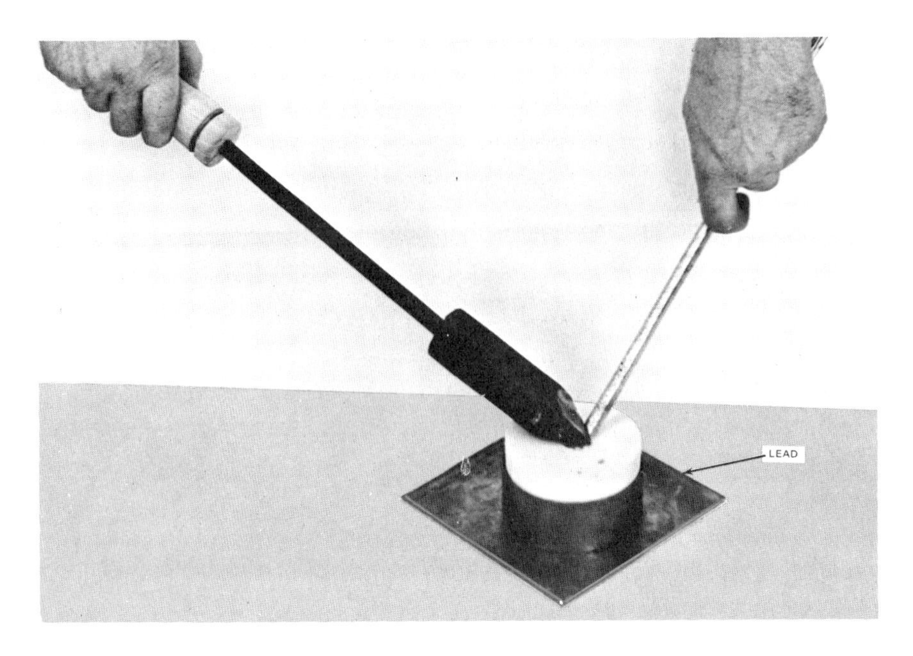

Figure 24-1 Tinning with sal-ammoniac

HOW TO SOLDER

1. Position the pieces to be soldered on a bench near the furnace.
2. Apply the proper flux to the surfaces to be soldered. (See unit 22.)
3. Remove the tinned copper from the furnace. Plunge it into a dipping solution or wipe it lightly and quickly with a cloth. A sharp, fast, sizzling sound as the copper contracts the solution indicates that the temperature is correct.
4. Position the copper on the joint to allow maximum heat transfer from the copper to the metal.
5. Apply solder to the copper.
6. Move the copper along the joint with an even flow of solder.

SOLDERING SEAMS

A lap seam must be tacked before soldering in order to hold the pieces in position. Tacking is not necessary to prepare a grooved seam for soldering because the seam is held together by the lock. A riveted seam is soldered without tacking since the rivets secure the work. Rivets should be soldered for watertight jobs. Properly soldered seams are shown in figure 24-2.

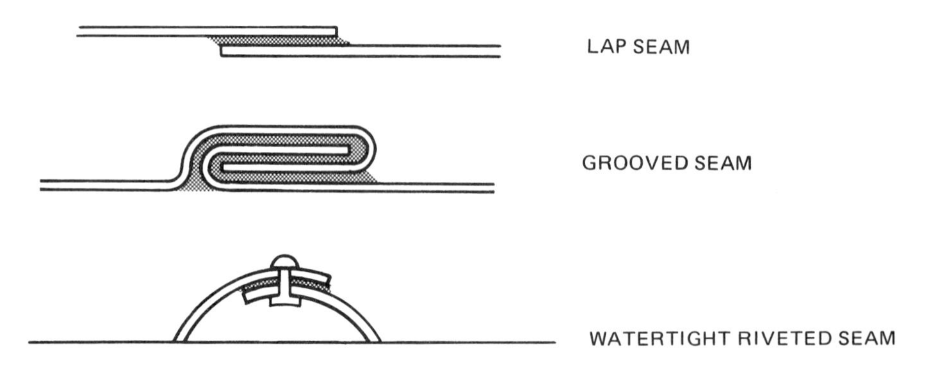

Figure 24-2 Soldered seams.

HOW TO SOLDER A LAP SEAM

1. Select the proper soldering copper.
2. Light the gas furnace and if necessary, prepare the soldering copper. Tin the copper.
3. Select the proper flux.
4. Place the tinned soldering copper in the fire pot or furnace.
5. Place the job to be soldered in the proper position on a suitable support, figure 24-3.

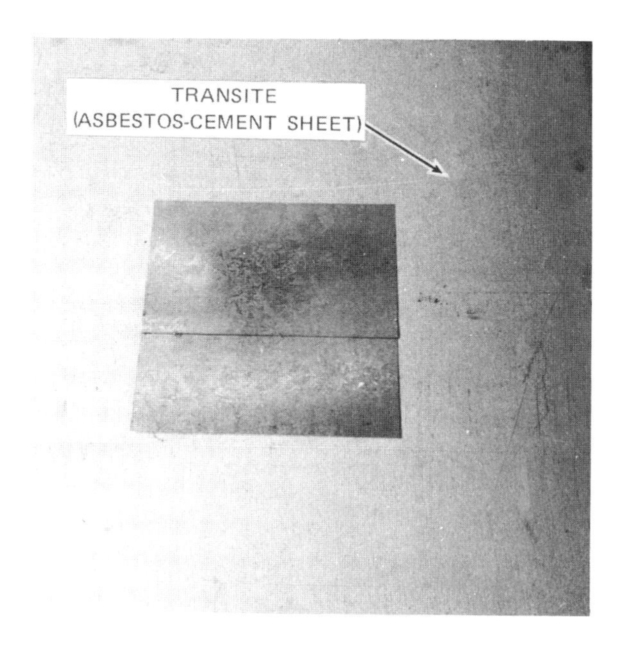

Figure 24-3 Job in place for soldering.

6. Dip an acid swab in the flux and apply it to the seam with one or two strokes.
7. Remove the soldering copper from the furnace and clean the tinned surface by dipping the end of the soldering copper in the dipping solution or by wiping off the tip with a damp cloth.
8. Pick up solder with the soldering copper as shown in figure 24-4.
9. Hold the copper at one end of the seam until the heat from the copper penetrates the metal.

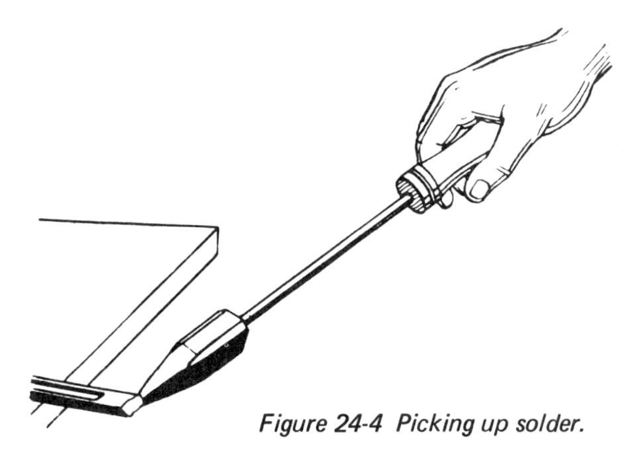

Figure 24-4 Picking up solder.

Figure 24-5 Tacking a seam.

Figure 24-6 Soldering a seam.

The metal being soldered must absorb enough heat from the soldering copper to melt the solder or the solder will not adhere to the metal. The bar of solder or a piece of steel or wood is used to hold the lap together while tacking. This makes a better job and eliminates the possibility of bad burns.

10. Tack the seam as often as is necessary to hold the pieces in position, figure 24-5.
11. Starting at one end of the seam, hold the soldering copper with a tapered side of the head flat along the seam, until the solder starts to flow freely into the seam.
12. Draw the copper very slowly towards you along the seam and add as much solder as necessary without raising the soldering copper from the job, figure 24-6.
13. Heat the soldering copper as often as it is necessary. The best job is produced if the seam can be started and finished without removing the soldering copper from the surface to be soldered.

If raw or boiled acid is used as a flux, wipe the excess acid off the seam with a clean damp cloth.

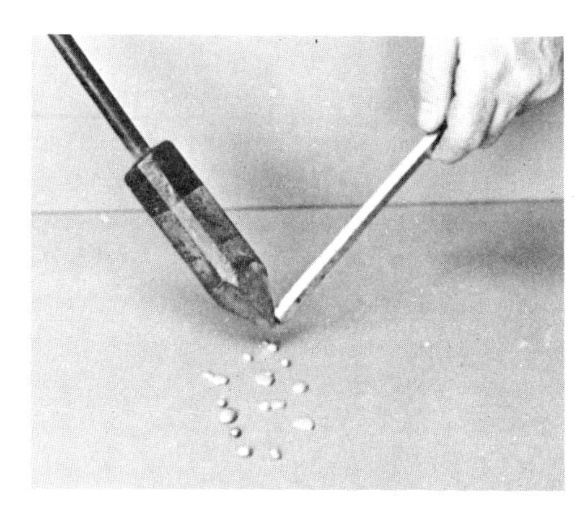

Figure 24-7 Making solder beads.

HOW TO SOLDER SQUARE, RECTANGULAR OR CYLINDRICAL BOTTOMS

1. Select the flux and a properly forged, tinned bottom copper.
2. Make solder beads (sometimes called shots) by holding the solder against the hot copper and allowing the beads to drop on the bench, figure 24-7.
3. Dip the acid swab in the flux and apply it to the seam with one or two strokes.
4. Place one of the cold beads of solder in the bottom of the job.
5. Remove the soldering copper from the gas furnace, clean it in the dipping solution, and place it in the bottom of the job as shown in figure 24-8.

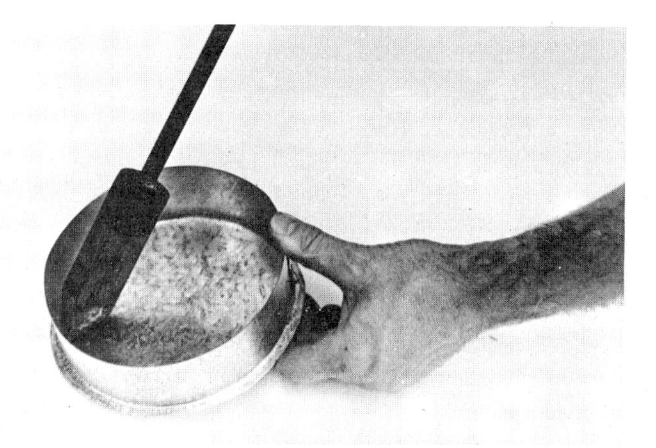

Figure 24-8 Soldering a bottom.

6. Hold the soldering copper in a stationery position until the solder starts to flow freely into the seam.
7. Draw the soldering copper very slowly along the seam, rolling the job on the edge of the bottom at the same time.
8. Add more solder beads as needed and reheat the copper as often as necessary.

SUMMARY REVIEW

A. Place the answers to the following questions in the column to the right.

1. How does soldering differ from welding? 1. ______________________
2. List three methods of soldering. 2. ______________________
____
____
3. What cleaning agent is used on the hot copper for tinning? 3. ______________________
4. How does the worker know that the copper is hot enough to clean? 4. ______________________
5. What part of the soldering copper should be coated with solder for tinning? 5. ______________________

B. Insert the correct word in each of the following sentences.

1. A brilliant ________ flame in the furnace at the tip of the soldering copper indicates that the ________ is being ____________ off.
2. A properly heated soldering copper ____________ when plunged into the dipping solution.
3. The main purpose of a heated soldering copper is to ________________ heat.

C. Underline the correct word in each of the following sentences.

1. The heated soldering copper tip is (cleaned, forged, shaped) on a sal ammoniac block.
2. In soldering a bottom in a container or pan, the solder is applied in (bar, wire, bead) form.

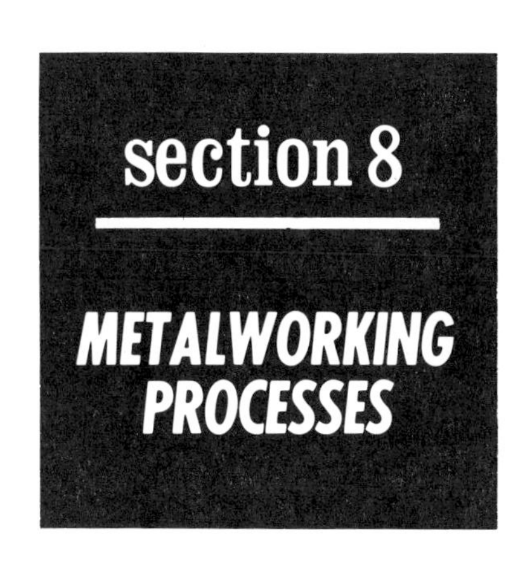

UNIT 25 PATTERNS

OBJECTIVES

After studying this unit, the student should be able to

- Identify the tools necessary for transferring a pattern.
- Compute the number of pieces obtainable from a sheet of metal.
- Transfer a pattern properly.

A pattern is the *stretchout* or *layout* of a job. It may be made from a blueprint or a shop ticket, figure 25-1. The shop ticket is usually a plan view and/or an elevation view of an object drawn either free-hand or with instruments. It is made up by the designer or shop foreman and given to the sheet metal worker for fabrication. The pattern used by the worker may be made of paper or metal.

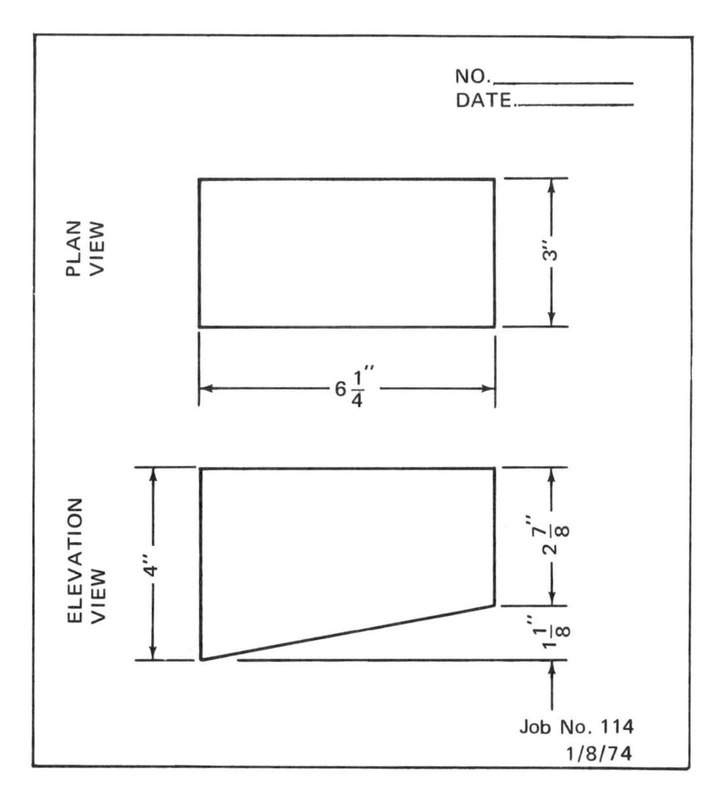

Figure 25-1 Sample shop ticket.

If the job is complicated and therefore requires preliminary layout, a paper pattern is made and transferred to metal stock. Then the job is cut out and formed. A paper pattern is made from drawing paper and is used only once. The metal stock to which it is transferred serves as the pattern for additional pieces or jobs.

If two or more pieces are required, trade practice is to lay out the job directly on metal. The metal layout is cut and used as a pattern to mark the remaining pieces. When sheet metal shops have repeated calls for jobs of the same kind, the patterns are stored to be used as needed. Metal patterns will retain their accuracy for an indefinite period of time if they are used carefully. They are sometimes called *master patterns.*

TYPES OF PATTERNS

There are three kinds of patterns: full, half, and pieced. The full pattern is an exact reproduction of the job to be done, figure 25-2. The half pattern represents one-half of a symmetrical job. This pattern must be transferred, turned over, and then transferred again to obtain the full stretchout, figure 25-3. Large or complicated jobs are made of two or more pieces. In many cases these pieces are exact duplicates so that one pattern can be made and then transferred to obtain the rest of the pieces, figure 25-4.

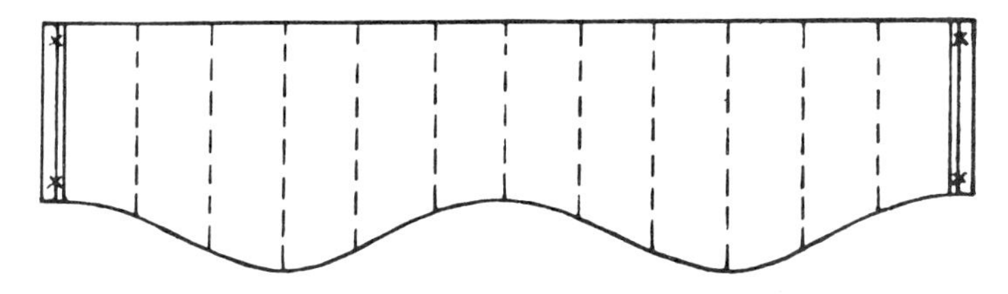

Figure 25-2 Full pattern.

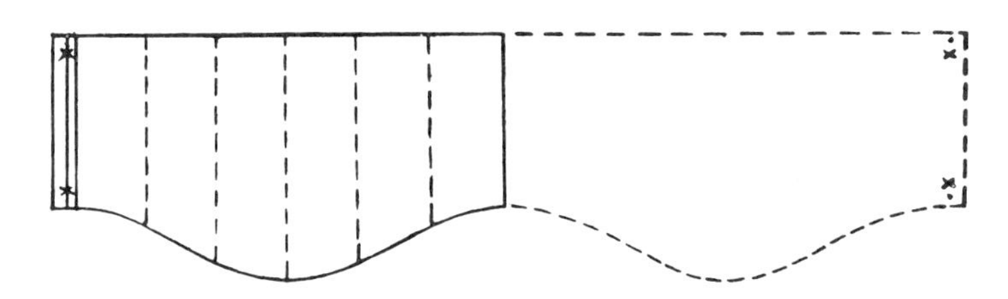

Figure 25-3 Half pattern.

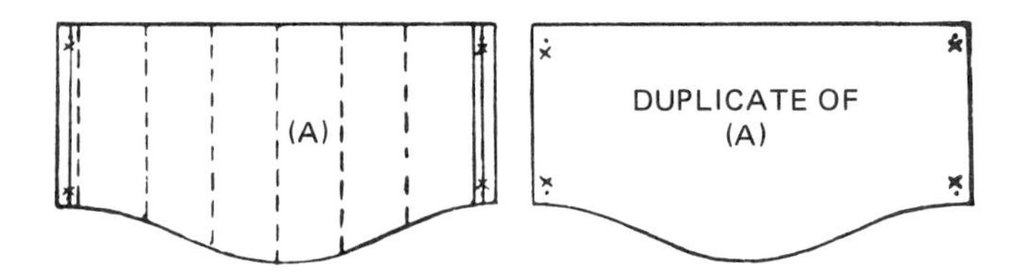

Figure 25-4 Pieced patterns.

HOW TO TRANSFER PATTERNS

A. Full Patterns

1. Check the pattern with the shop ticket for accuracy in transferring. Mistakes are often magnified in transferring the pattern.

2. Obtain a sharp prick punch and hammer to mark brake points and hole centers. Occasionally, patterns are outlined on the metal with close prick punch marks.

3. Sharpen a scratch awl or scriber to a long conical point for scribing outlines on ordinary metal.

When marking special metals, such as aluminum, Monel®, or stainless steel, use a hard pencil (9H) to avoid scratches that would weaken or ruin the finish.

4. Obtain snips or scissors, iron weights, C-clamps, and stock.
5. Cut out the outline of the pattern accurately with scissors or snips.
6. Place the sheet of metal to be used on a wooden bench or on a flat surface.

An iron surface should not be used since the prick punch would become dull or would break.

7. Place the pattern on the sheet of metal in the most economical position, figure 25-5. In most cases, the pattern is placed at one corner of the sheet.

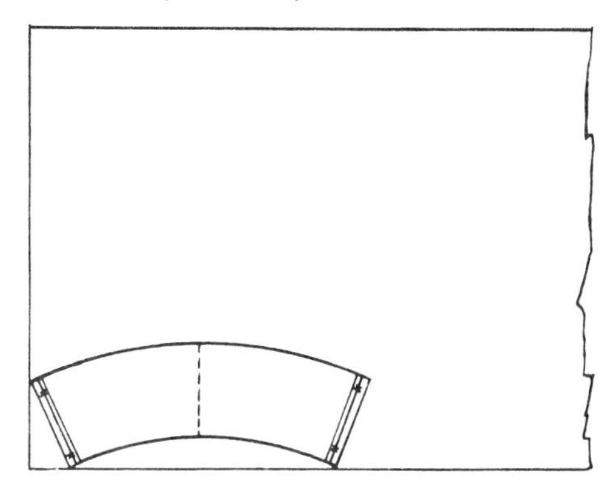

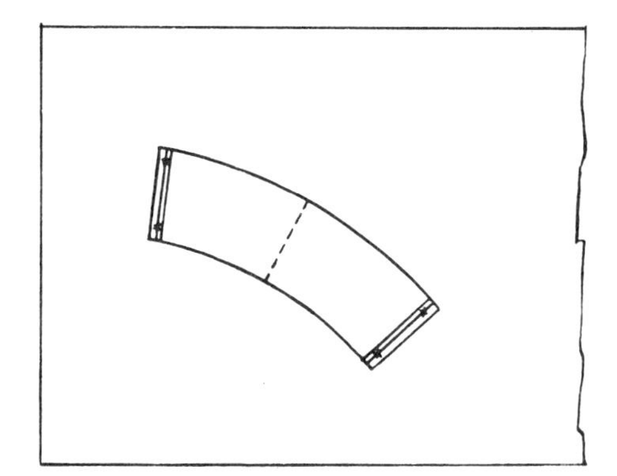

Figure 25-5 Placement of paper pattern on stock.

8. Put weights on the pattern to hold it in place, figure 25-6.

   The heads of the double-seaming stakes, square blocks of steel, or short pieces of railroad rail make good weights to hold a pattern in place.

   or

   Clamp the pattern in place with C-clamps, figure 25-7. For patterns involving many details and accurate fitting, the C-clamp method is the best. In most cases, the weight method is sufficient.

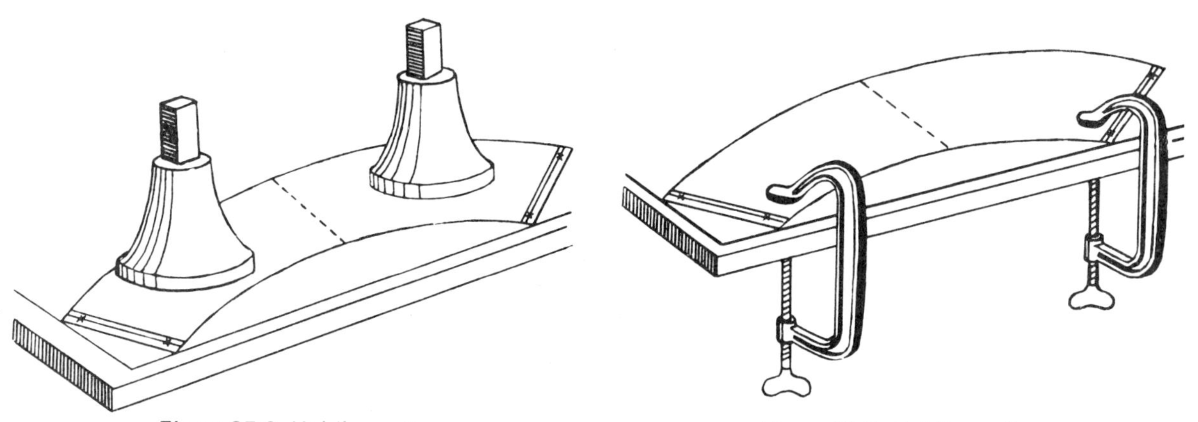

Figure 25-6 Holding pattern in place with weights.

Figure 25-7 Holding pattern in place with c-clamps.

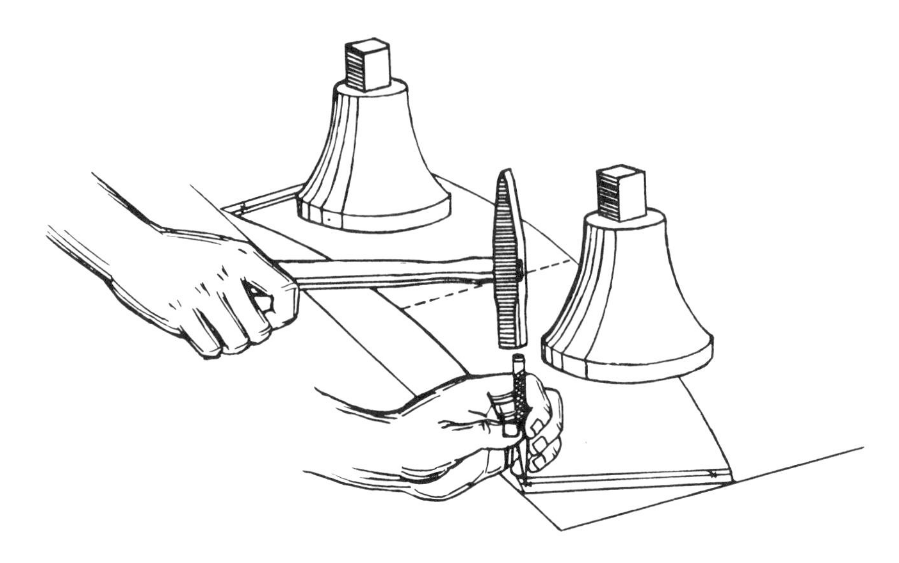

Figure 25-8 Prick punching brake lines.

9. With a sharp prick punch and hammer, lightly mark the brake lines and the centers of holes to be drilled or punched, figure 25-8.

Always use a sharp prick punch. A dull punch enlarges the marks on the pattern and eventually spoils the pattern for accurate work.

10. Hold the edges of the pattern in place with the left hand and scribe the outline of the pattern with a sharp scratch awl or scriber held in the right hand, figure 25-9. To obtain a true duplicate of the pattern, tilt the scratch awl with the point towards the pattern when scribing the line, as shown in figure 25-9.

Use a sharp scratch awl with a long tapered point for accurate work.

11. Remove the weights or C-clamps and cut out the metal. When more than one piece is wanted, the metal is cut out and used as the pattern for the rest of the pieces.

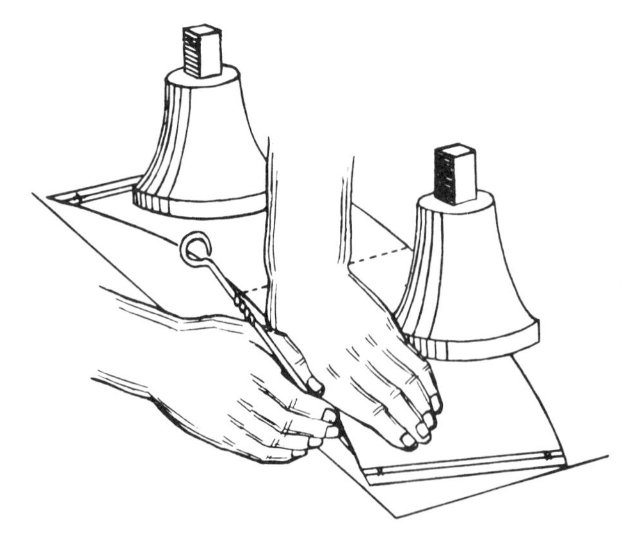

Figure 25-9 Scribing the outline of the pattern.

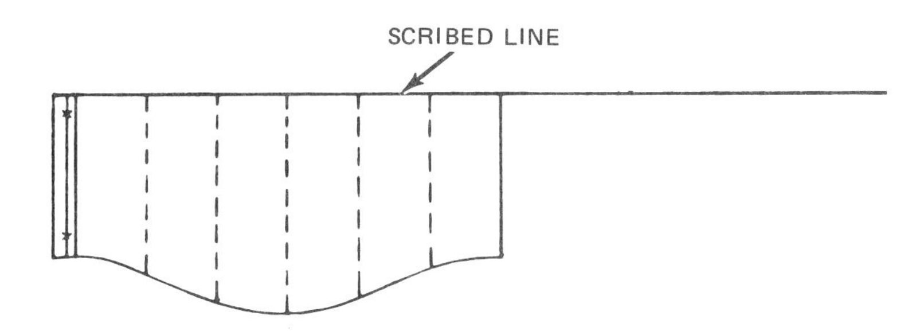

Figure 25-10 Half pattern.

B. *Half Patterns*

1. Cut out the outline of the half pattern accurately with the snips or scissors.
2. Place the sheet of metal to be used on a wooden bench or on a flat surface.
3. Using a straightedge, scribe a line near the edge of the sheet with a sharp scratch awl or scriber.
4. Place the half pattern on the sheet of metal with one side along the scribed line as in figure 25-10. Hold the pattern in place with weights or C-clamps.
5. Mark the brake lines lightly with a sharp prick punch and hammer.
6. Scribe the outline of the half pattern with a sharp scratch awl or scriber.
7. Remove the weights or C-clamps.
8. Turn the half pattern over like the leaf of a book.
9. Place the half pattern with the edges flush with the two lines as in figure 25-11.

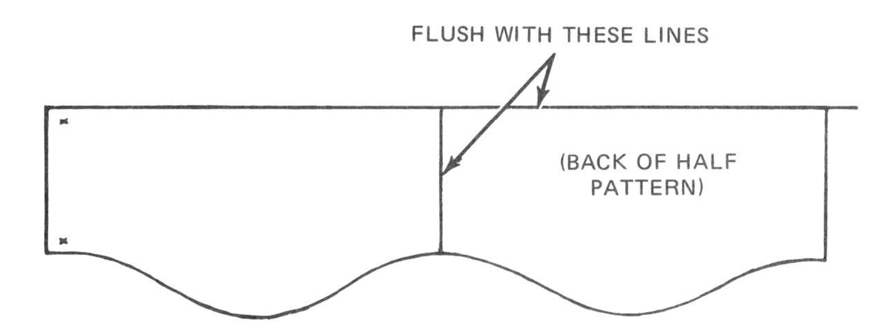

Figure 25-11 Full pattern.

10. Replace the weights or C-clamps on the pattern.
11. Mark the brake lines with a sharp prick punch and hammer.
12. Scribe the outline of the half pattern with a sharp scratch awl or scriber.
13. Remove the weights or C-clamps.
14. Cut out the metal accurately to complete the full pattern. When more than one piece is wanted, the full pattern is used to mark the rest of the pieces.

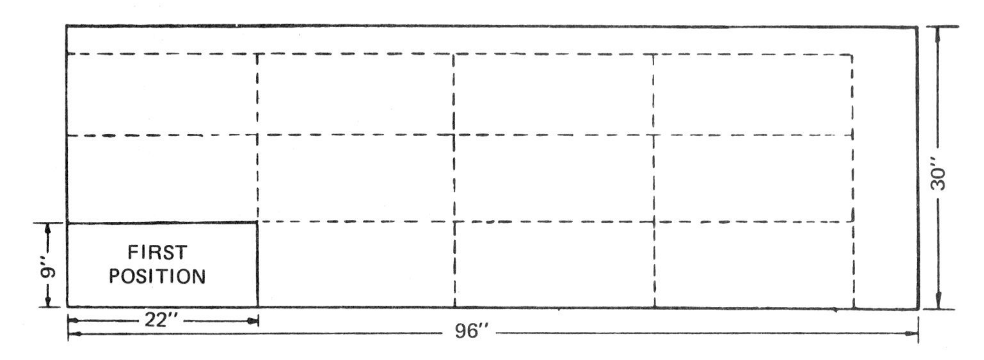

Figure 25-12 First position of pattern on sheet.

HOW TO FIND NUMBER OF PIECES OBTAINABLE

A. *Computation*

1. Make a rough sketch of the sheet, including the shape of the pattern in one corner, figure 25-12. Indicate the dimensions of the sheet and the pattern on the sketch.

2. Divide the width of the pattern into the width of the sheet and drop the fractional remainder.

$$\frac{30}{9} = 3 \text{ pieces } \underline{\text{Ans.}}$$

3. Divide the length of the pattern into the width of the sheet and drop the fractional remainder.

$$\frac{96}{22} = 4 \text{ pieces } \underline{\text{Ans.}}$$

4. Multiply these two numbers. $3 \times 4 = 12$ pieces <u>Ans.</u>

5. Make another sketch of the sheet in which the position of the pattern is rotated 90 degrees, figure 25-13.

6. Calculate the number of pieces which can be cut from the sheet in this position.

Width: $\frac{30}{22}$ inches = 1 piece Length: $\frac{96}{9}$ inches = 10 pieces

Total pieces: $1 \times 10 = 10$ pieces, <u>Ans.</u>

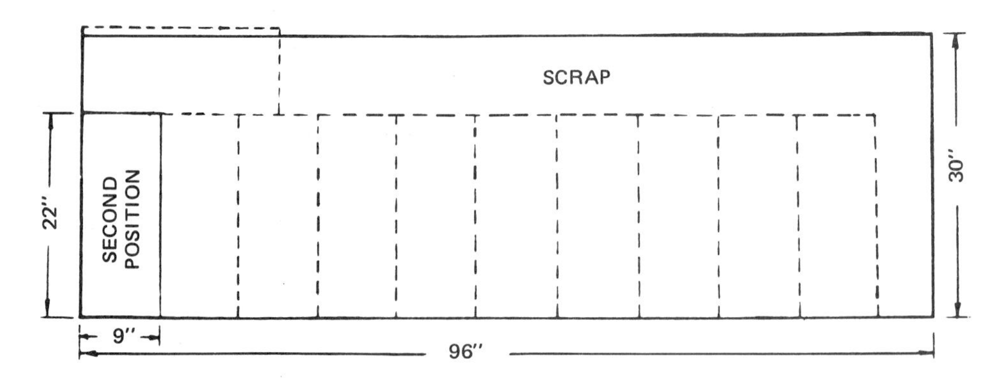

Figure 25-13 Second position of pattern on sheet.

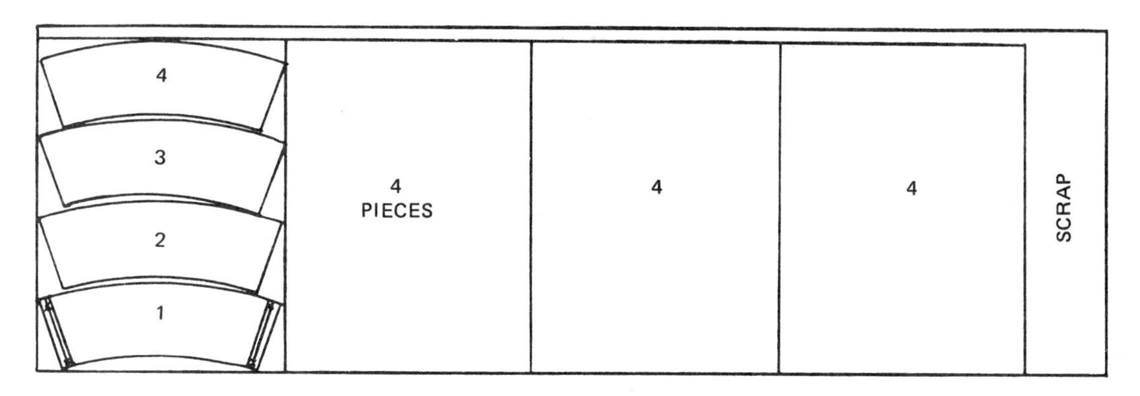

Figure 25-14 First method of placing pattern.

7. Therefore, the first position gives the greatest number of pieces, 12, and should be used, as shown in figure 25-12.

B. *Layout*

1. Place the pattern on the sheet of metal at the lower left-hand corner and mark around it with a pencil, figure 25-14, position 1.
2. Move the pattern in successive steps along the width of the sheet to find how many pieces can be obtained. Figure 25-14 shows that four can be obtained.
3. Measure the maximum length of the pattern and mark this length on the sheet as many times as possible. Figure 25-14 shows four times.
4. Multiply these two numbers. 4 x 4 = 16 pieces.
5. Rotate the pattern 90 degrees and place it on the sheet of metal at the lower left hand corner as in figure 25-15, position 1.
6. Move the pattern along the length of the sheet to find out how many pieces can be cut from the length of the sheet.

The pattern can be nested in groups. Figure 25-14 shows the pattern stepped off four times horizontally and this distance stepped off the length of the sheet three times. The rest of the sheet should be checked to find if any more pieces can be obtained:

4 x 3 = 12 pieces. The position in figure 25-14 gives the larger number of pieces (16).

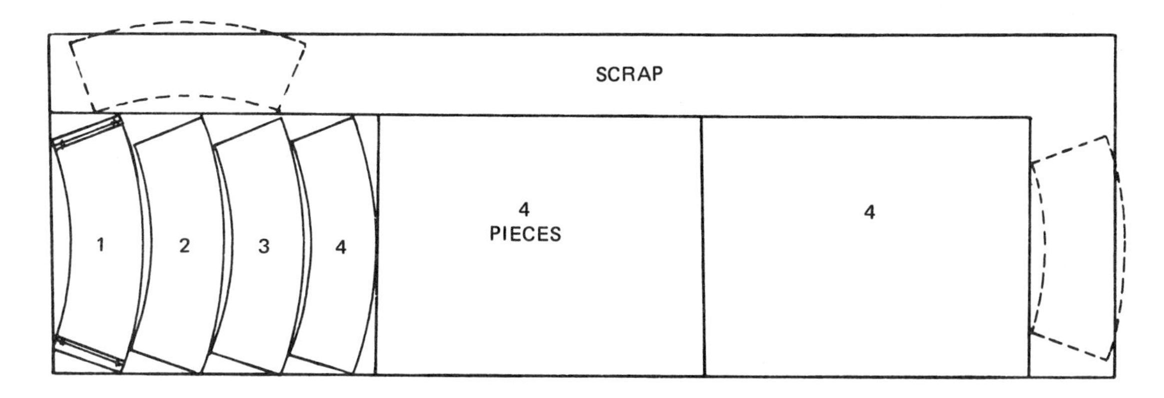

Figure 25-15 Second method of placing pattern.

SUMMARY REVIEW

A. Place the answers to the following questions in the column to the right.

1. List the tools necessary to transfer and cut out a metal pattern on a galvanized iron sheet. 1. ______________________
____
____
____
____
____

2. List the two materials that patterns are usually made of. 2. ______________________
____

3. List the three types of patterns used in layout work. 3. ______________________
____
____

4. When metal patterns are stored for future use, what are they called? 4. ______________________

B. Insert the correct word in each of the following sentences.

1. A pattern is the __________ or __________ of the job.
2. A pattern of several pieces is usually placed directly on __________.
3. Metal patterns retain their __________ for an indefinite period.
4. A metal pattern should be cut with sharp __________ for accuracy.

C. Underline the correct word in each of the following sentences.

1. Patterns should be outlined on stainless steel with a (scratch awl, prick punch, 9H pencil).
2. Patterns are usually placed at one (corner, end, side) of the sheet to avoid waste.
3. Detailed patterns should be secured to the metal sheet with (screws, nails, clamps).
4. Using a pattern with overall dimensions of 8 1/2 x 30 1/2 inches, (10, 15, 12, 16, 17) pieces can be cut from a 36 x 96 inch sheet of 26-gage galvanized iron.

UNIT 26 *FLANGING*

OBJECTIVES

After studying this unit, the student should be able to

- List the purposes for a flange.
- List the methods of making a flange.
- Make an inside and outside flange properly.

A flange is an edge bent on a job, and it is used for stiffening or for making a connection to another piece, figure 26-1. Flanges on irregularly shaped jobs must be made by hand, while small flanges on the outside of cylinders can be made with the turning machine or the burring machine.

Figure 26-1 Flange.

Considerable skill is necessary to do a good flanging job. Since flanging involves working the metal, it is desirable not to strike the flange any more often than necessary to prevent the metal from cracking. The flange should be made as small as possible because the larger the flange, the more times it will have to be struck and the greater the chances are of cracking it.

STRETCHING AND SHRINKING METAL

When making a flange on the outside of a job, the metal must be stretched. When making a flange on the inside of a job, the metal must be shrunk.

When an edge is turned outward on a cylinder, the diameter and circumference of the outer part of the edge increases, figure 26-2. Diameter *A* increases to diameter *B;* The edge at *D* is stretched from *C.* As the metal stretches, it becomes thinner at the outer edge: note that *F* is thinner than *E.* The metal should be flattened and bent over with a hammer at the same time. This thins out the metal gradually and prevents cracks from developing.

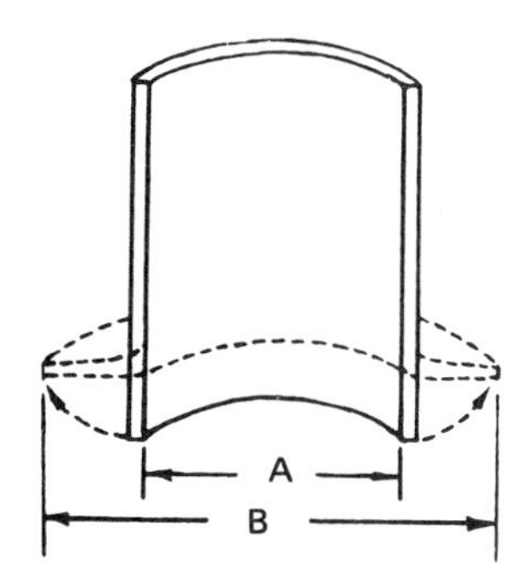

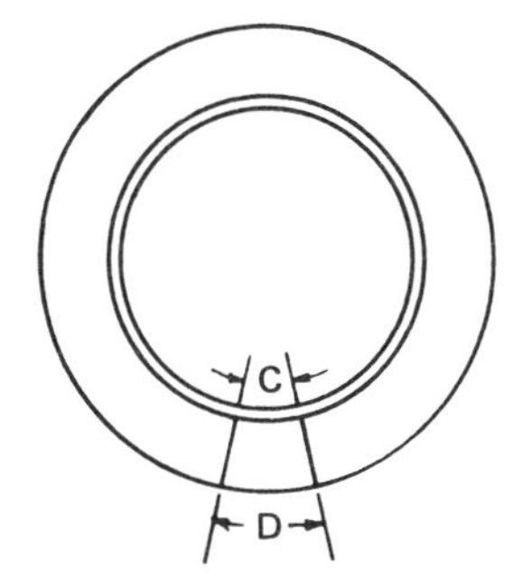

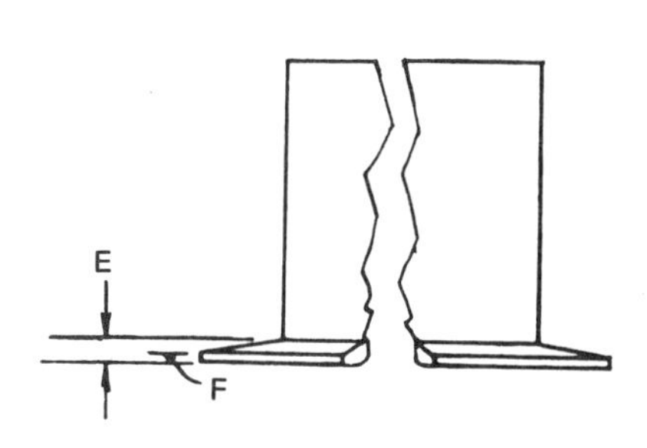

Figure 26-2 Outward flange: metal is stretched.

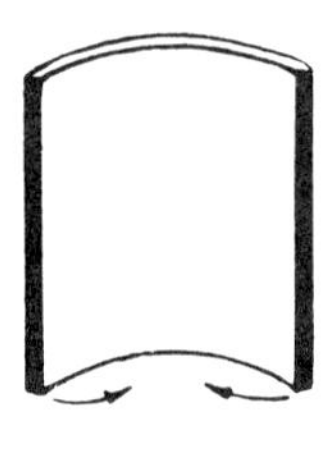

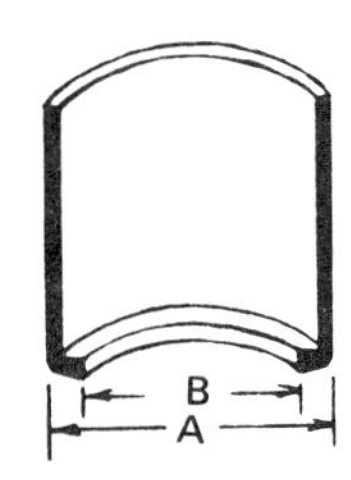

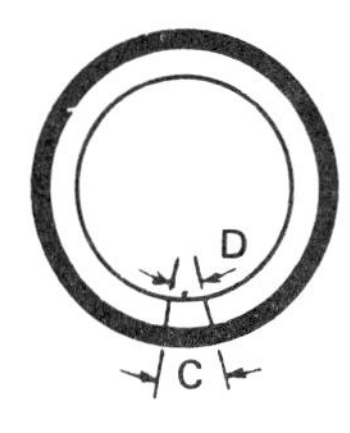

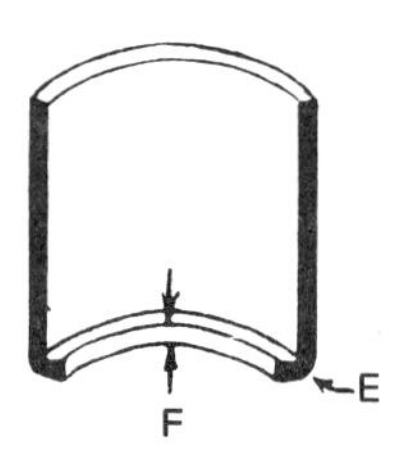

Figure 26-3 Inward flange: metal is shrunk.

When turning an edge inward, the metal must be shrunk, figure 26-3. The outer part of this edge is now reduced in diameter and circumference. Diameter *B* is less than diameter *A*. The edge at *C* is reduced to *D*. Since the amount of metal remains the same, the edge must become thicker; note that *F* is thicker than *E*. This is called shrinking the metal and is usually done with a mallet. Since the metal is made thicker while pounding and turning the edge at the same time, care must be taken to prevent the metal from being stretched and the edge from buckling and becoming wavy.

CONCLUSIONS

- Since heavy metal will stretch more than light metal without becoming too thin, a wider outside flange can be made on heavy metal than on light metal.
- When making an inside flange on a cylinder, the metal must be reduced in circumference. Therefore, a narrow flange should be made when shrinking metal because a wide flange is apt to buckle.
- Since striking metal makes it hard and brittle, it should not be struck any more than necessary when flanging.
- A mallet is less apt to stretch metal and so is usually used when shrinking metal.

HOW TO FLANGE AN OUTSIDE EDGE ON A STAKE

1. Mark the width of the flange on the inside of the job with a marking gage.
2. Place the square stake in a hole in the bench plate. Any stake or plate with a square edge can be used.
3. Hold the job with the edge to be flanged on the stake and strike the edge with the peen end of the riveting hammer, figure 26-4.

Figure 26-4 Starting the flange.

Figure 26-5 Smoothing the flange.

4. While striking the edge, turn the job until the entire edge to be flanged has been started. As the metal stretches, lower the job and keep striking the flange with the peen of the hammer until the required angle is obtained.
5. Place the conductor stake in a hole in the bench plate.
6. Hold the flange on the surface of the conductor stake and smooth the flange with the flat face of the hammer, figure 26-5.

When forming a 90-degree flange, the job is smoothed several times during the flanging operation. T-joints are flanged in a similar manner, figure 26-6. The amount of flanging to be done depends upon the types of T-joint.

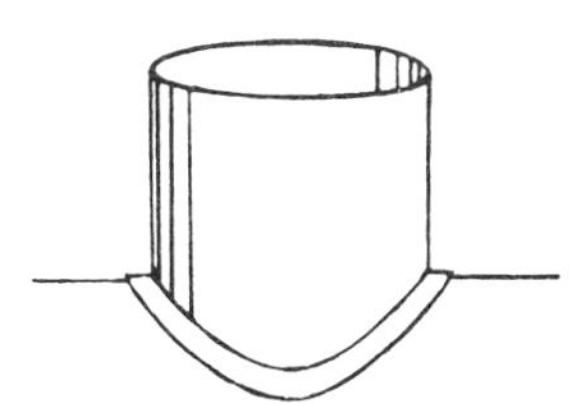

Figure 26-6 Flanged T-joint.

HOW TO SHRINK AN INSIDE EDGE ON A STAKE

1. Mark the width of the flange on the outside of the job with a marking gage.
2. Place the conductor stake in a hole in the bench plate. Any round stake of suitable curvature can be used.
3. Hold the job with the edge to be shrunk on the end of the stake and strike the edge to be shrunk with a mallet, figure 26-7.

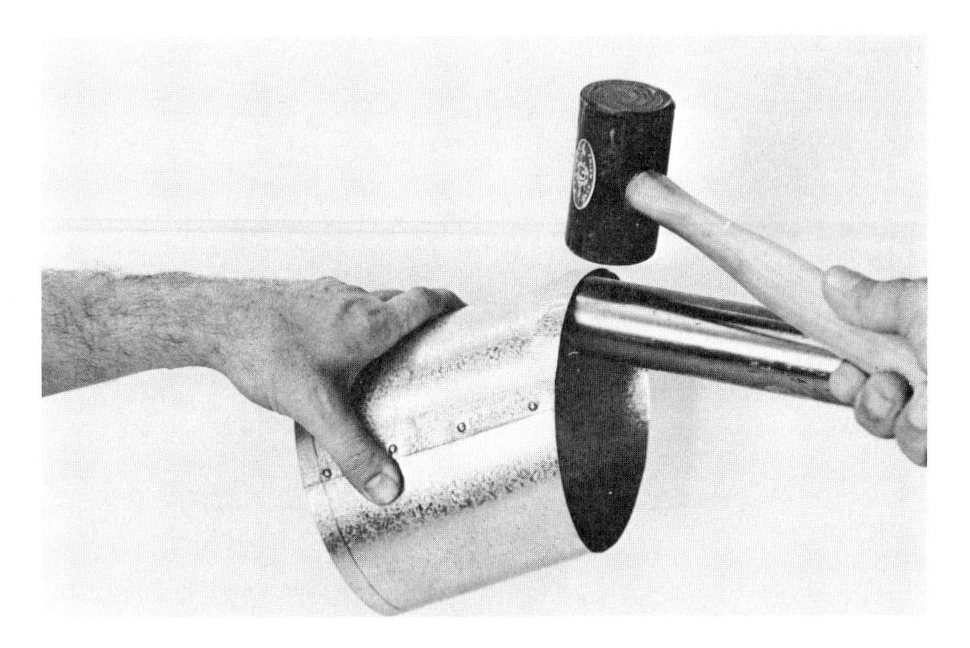

Figure 26-7 Shrinking with a mallet.

4. While striking the edge, turn the job until the entire edge to be shrunk has been started. As the metal shrinks, lower the job and keep striking the edge until the required angle is obtained.

HOW TO FLANGE AN EDGE USING A DOLLY

1. Set the gage on the turning machine for the required width of the flange and form the edge as near as possible to the angle needed.
2. Place the job on a bench plate or any flat surface to finish the outer edge of the job. Tap the edge of the flange with hammer until the job lies flat on the bench plate.
3. Hold the dolly in the left hand with the rounded edge of the dolly flush with the inside edge of the flange and tap the outside edge with the flat face of a small hammer, figure 26-8. Continue tapping the edge with the hammer while moving the dolly back and forth until the edge is at the angle required.

Flanges should be free of buckles when finished.

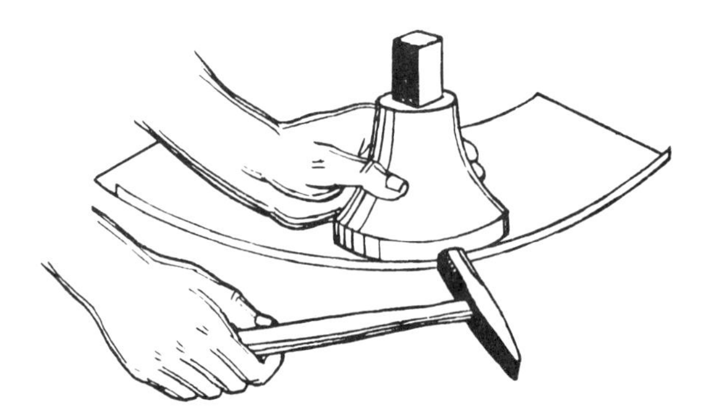

Figure 26-8 Flanging the heel.

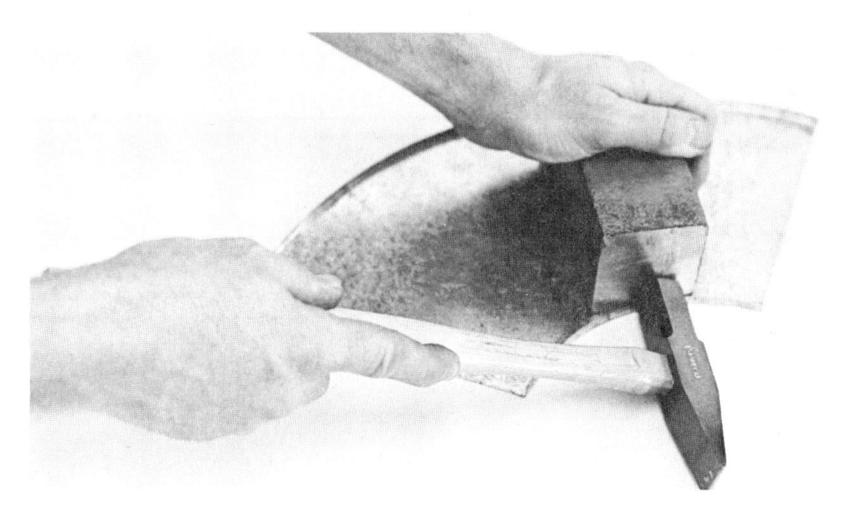

Figure 26-9 Flanging the throat.

4. Flange the other edge of the job in the same way, using the flat side of the dolly to back up the edge, figure 26-9.
5. Bulging is caused by not shrinking the outside of the flange enough. If the job is wavy and bulges up, it can be straightened by tapping the flanged edge, figure 26-10. If the job is bulged down, it can be straightened by turning the job over so that the flanged edge is on the bench plate and tapping the bulge until it is straightened, figure 26-11.

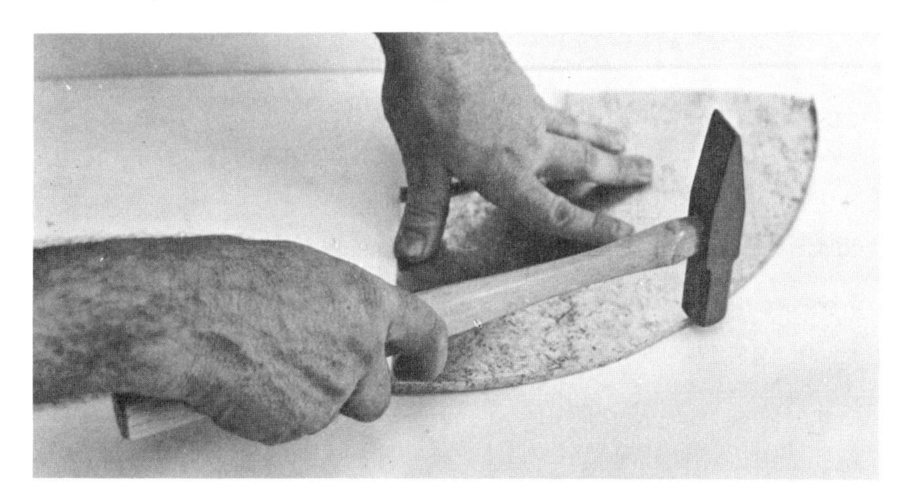

Figure 26-10
Flange bulged up.

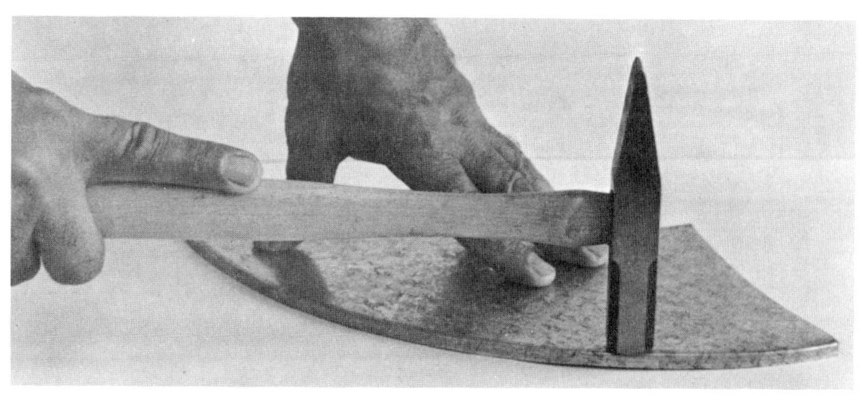

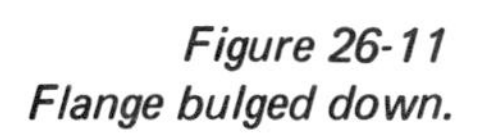

Figure 26-11
Flange bulged down.

SUMMARY REVIEW

A. Place the answers to the following questions in the column to the right.

1. State two general purposes of a flange. 1. ______________________
   ____
2. State two methods of making a flange. 2. ______________________
   ____
3. State the two general types of flanges. 3. ______________________
   ____
4. What may happen to a metal if it is hammered repeatedly? 4. ______________________
   ____
5. What is a flange? 5. ______________________

B. Insert the correct word in each of the following sentences.

1. When making an outside flange on a pipe, the metal is ____________ .
2. When making an inside flange the metal is ____________ .
3. A flange should be made as ____________ as possible to prevent cracking.
4. If a large flange is made, it may cause a ____________ .
5. Cracks can be prevented during flanging by ____________ and ____________ the flange at the same time.

C. Underline the correct word in each of the following sentences.

1. When flanging to the outside, the circumference (decreases, increases, remains the same).
2. A (hammer, mallet, lock former) is used in forming an outside flange.
3. In smoothing out a 90-degree flange, the (face, peen, edge) of a hammer is used.
4. A (hand dolly, hatchet stake, rivet set) is used in flanging a curved or irregular job.
5. While peening a flange, the metal edge must be kept (flat, upright, on edge) on a stake.

UNIT 27 SINGLE SEAMS

OBJECTIVES

After studying this unit, the student should be able to

- State the uses of the single seam.
- Determine the allowance for a single seam.
- Make a single seam properly.

The single seam is a folded seam, standing at right angles to the surface of a job, figure 27-1. It not only acts as a seam, but it may also act as a stop for covers. This seam is often referred to as a *peened seam.*

Figure 27-1 Single seam.

The single seam can be made by hand or machine. The hand method is used in shops when a setting-down machine is not available and when large seams are wanted since the capacity of setting-down machines is usually limited.

Three precautions must be observed to produce a workmanlike single seam:

- The allowances for the edges must be correctly distributed.
- The edges should be cut accurately and formed evenly.
- The setting hammer must be handled with care so that the job is not marred.

USES

A single seam is a combination stop and collar for attaching ends, tops, and bottoms to cylinders and boxes. It may be used when the job does not have to be watertight and when little strength is needed. This seam is also used on round, adjustable elbow fittings.

Figure 27-2 shows a single seam used on a cover in which a collar is attached to the top. When using this seam as a stop to prevent the cover from sliding into the container, the hem is formed on the inside of the collar.

Figure 27-2 Cover with collar.

EDGE PREPARATION

The edges for a single seam are prepared as shown in figure 27-3, and then one edge is forced or peened over the other. This operation is called *setting down.* On square or rectangular jobs, the single seam is either soldered or riveted to hold the edges in place.

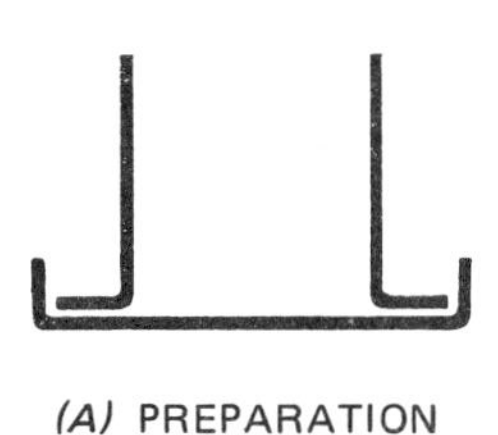

(A) PREPARATION

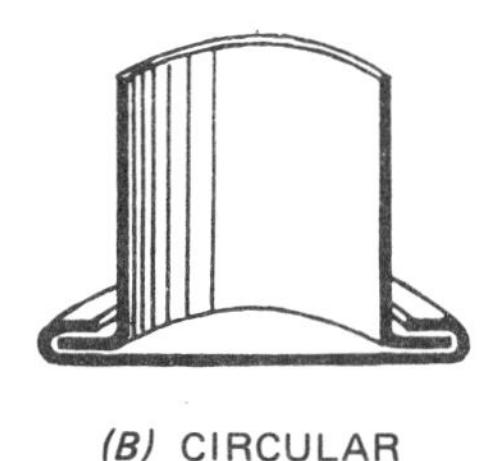

(B) CIRCULAR

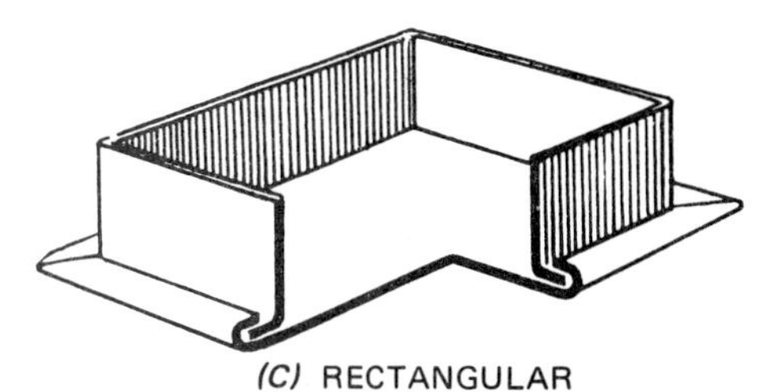

(C) RECTANGULAR

Figure 27-3 Setting down the seam.

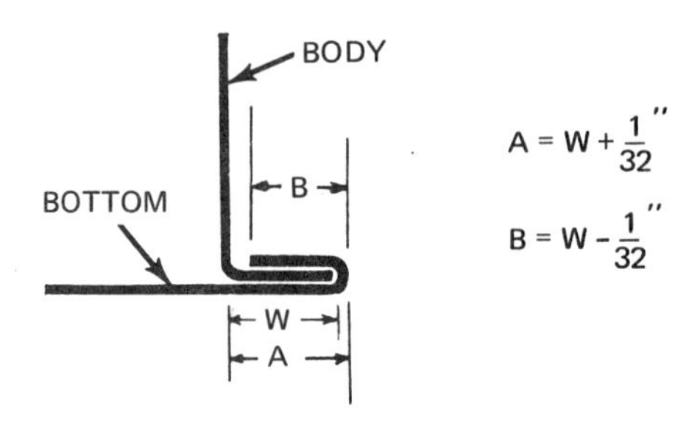

Figure 27-4 Allowances.

ALLOWANCES

The allowance of metal on the body of the job is equal to the width *W*, figure 27-4. The allowance on one side of the other piece is equal to 2W but is so distributed that *A* is 1/32 inch larger than *W*, and *B* is 1/32 inch smaller than *W*. This distribution avoids interference of the folded edge with the body of the job.

Example: Find the correct distribution for the bottom of a job if the allowance on the body is 1/8 inch.

Solution: W = 1/8

A = 1/8 + 1/32 B = 1/8 - 1/32

A = 5/32 inch, *Answer* B = 3/32 inch, *Answer*

HOW TO CONNECT A BOTTOM TO A CYLINDER

1. Burr the edge on the cylinder and the bottom to the required allowances, using the burring machine, figure 27-5A and 27-5B.
2. Snap the bottom over the burred edge of the body, figure 27-5C.
3. Rest the bottom on a flat stake or plate with one edge flush with the right-hand side of the stake or plate.

By keeping the edge flush with the edge of the plate or stake, it is possible to bring the burred edge of the bottom over the cylinder without hitting the plate.

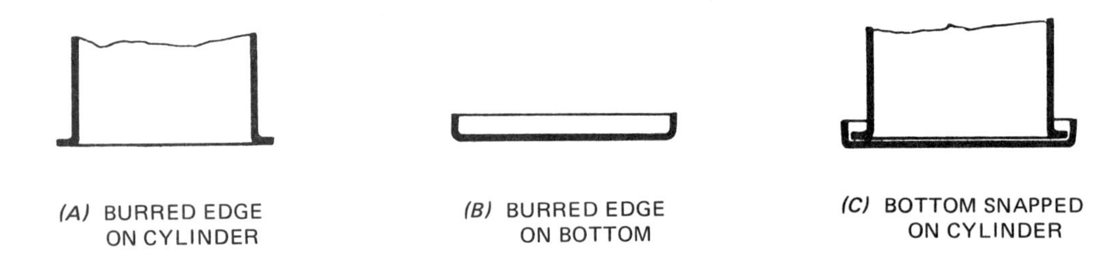

(A) BURRED EDGE ON CYLINDER *(B)* BURRED EDGE ON BOTTOM *(C)* BOTTOM SNAPPED ON CYLINDER

Figure 27-5 Connecting a bottom and cylinder.

4. Using the face of the setting hammer, bend the burred edge of the bottom over the burred edge of the cylinder to approximately a 45-degree angle, figure 27-6. Bend the edge of the bottom gradually with the face of the setting hammer until the edge is flat. The burred edge is bent gradually so the seam will not buckle or come apart. The peen of the setting hammer can be used to finish the flattened edge.

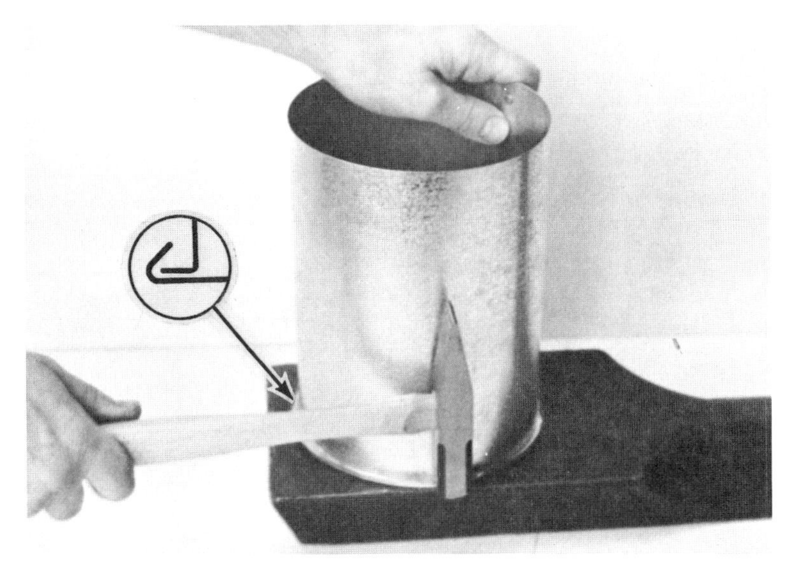

Figure 27-6
Bending the edge.

HOW TO CONNECT A COLLAR TO A PITCHED TOP

1. Burr the edge on the cover, and the collar to the required allowance, using the burring machine, figures 27-7A and 27-7B.
2. Snap the cover over the burred edge of the collar, figure 27-7C.

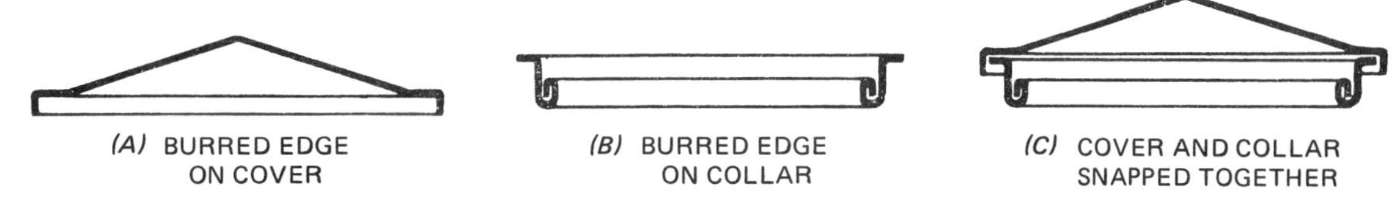

Figure 27-7 Connecting a collar and pitched top.

3. Hold the edge to be peened on the square head stake as shown in figure 27-8.
4. Peen the seam, following the same procedure as in Connecting a Bottom to a Cylinder. On most jobs of this kind the peen end instead of the face of the setting hammer is more convenient to use.

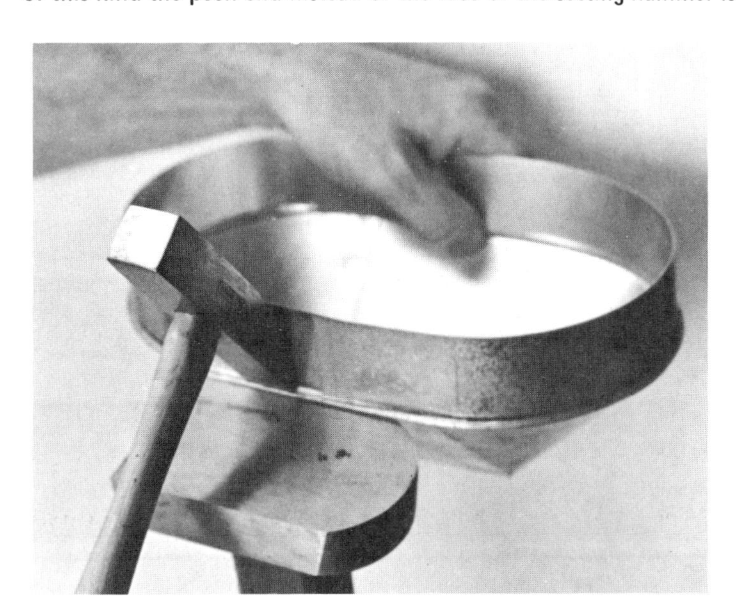

Figure 27-8
Peening the edge.

SUMMARY REVIEW

A. Place the answers to the following questions in the column to the right.

1. List three uses of the single seam. 1. ____________________

____

____

2. List two purposes of the single seam on a cover. 2. ____________________

____

3. On what type of fitting would a single seam be used? 3. ____________________

4. Briefly list the three main factors in making single seam. 4. ____________________

____

____

B. Insert the correct word in each of the following sentences.

1. In making a large single seam, the ____________ method is used.
2. Single seams formed on a setting down machine are ____________ the size and capacity of machine.
3. In setting the burred edge of a single seam, the bending is done gradually to prevent ____________.
4. The setting hammer must be used with care to avoid ____________ the job.

C. Underline the correct word in each of the following sentences.

1. The single seam is a (watertight, airtight, loose) connection.
2. A single seam can be used as a (flange, hem, stiffener) for added strength, although its strength is not as great as other seams.
3. Edges are prepared for a single seam on a small cylinder by using a (burring machine, mallet, setting hammer).
4. To make a single seam watertight, it must be (riveted, soldered, pressed).

UNIT 28 DOUBLE SEAMS

OBJECTIVES

After studying this unit, the student should be able to

- State the uses of the double seam.
- Calculate the proper allowances for a double seam.
- Double seam bottoms, corners, heels, and throats.

The double seam is a single seam bent over against the body of the job, figure 28-1. It is used to connect parts of the jobs where strength is required. The allowances for single and double seams are the same. It is especially important for the double seam allowance to be accurate, however, so that the seam is not crushed during fabrication.

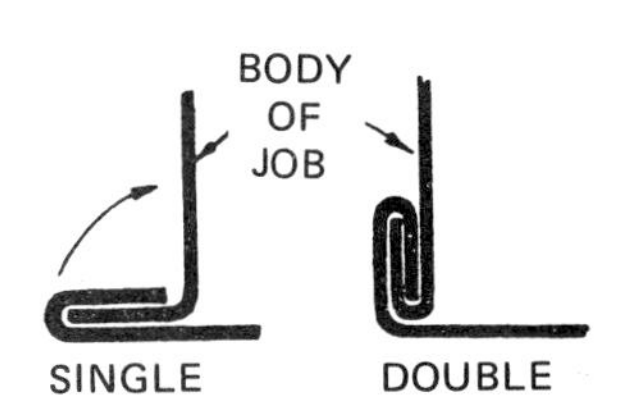

Figure 28-1 Seams.

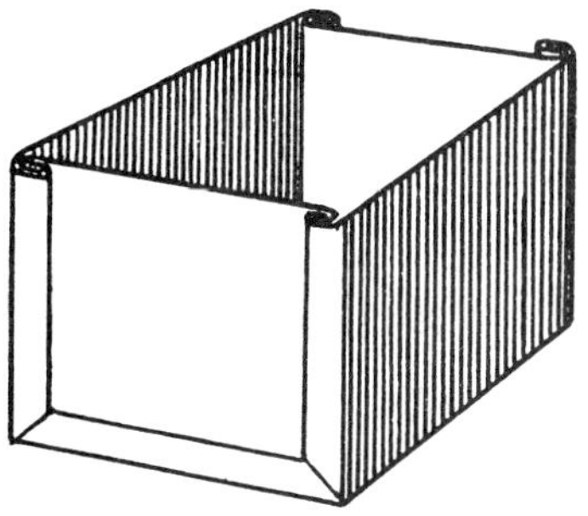

Figure 28-2 A double-seamed box.

The double seam can be finished by hand or machine. The hand method is more practical because the double-seaming machine is limited to a very narrow width of seam on tin plate and light metal. Tapering jobs are double seamed by hand.

For a double seam to be adequate, it must be made as narrow as possible, the edges must be cut accurately and formed evenly, and the job must be backed up by the proper stake or dolly during the forming and setting operations. Double seaming may be done on several types of stakes: the double-seaming stake, the double-seaming stake with four heads, the square bar stake, and special stakes.

USES

The double seam is used to attach parts of jobs together. It is a mechanically secure fastening and does not depend upon solder, rivets, or bolts to hold it in place. Double seams are used where strength is required and can be used on square, rectangular, or circular jobs such as hoppers, boxes, tanks, pails, buckets, and square or rectangular pipe work.

The seam is sufficiently airtight so that it is dustproof, and soldering is not necessary. If the job is to be watertight, however, the seam must be soldered. The double seam also makes a strong, stiff corner lock for square and rectangular pipe, elbows, and fittings.

ALLOWANCES

A double-seamed edge for circular jobs should be narrow, because the metal is shrunk when the edge is bent over. Therefore, a narrow seam is easier to form and makes a neater looking job. A wide edge buckles and appears wavy, figure 28-3.

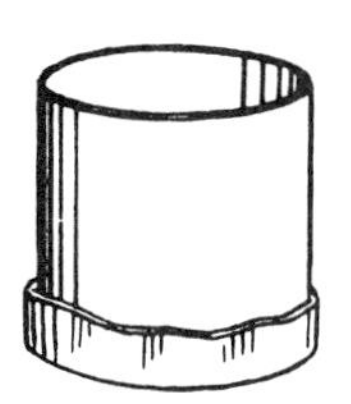

Figure 28-3 Wavy edge.

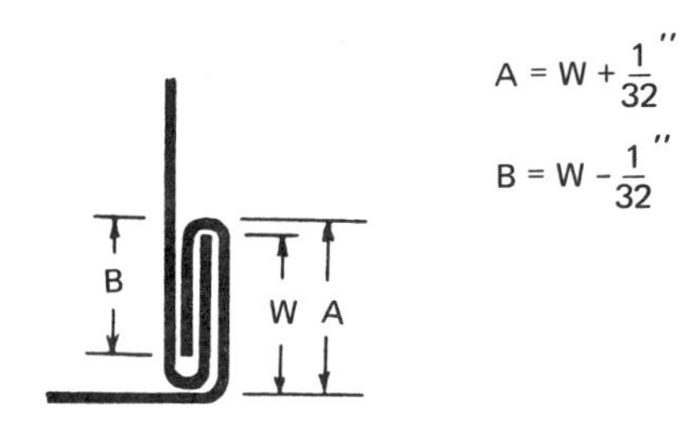

Figure 28-4 Allowances.

The allowance of metal on the body of the job is equal to the width *W*, figure 28-4. The allowance on one side of the other piece is equal to 2W but is so distributed that *A* is 1/32 larger than *W* and *B* is 1/32 smaller than *W*.

Example: Find the correct distribution for the bottom of a job if the allowance on the body is 3/16 inch.

Solution: $W = \frac{3}{16}$

$A = \frac{3}{16} + \frac{1}{32}$ $\qquad B = \frac{3}{16} - \frac{1}{32}$

$A = \frac{7}{32}$ inch, *Answer* $\qquad B = \frac{5}{32}$ inch, *Answer*

If the inside edge is equal to or greater than *W*, the edge interferes with the body of the job and the double seam crushes, figure 28-5.

Figure 28-5 Interference.

EDGE PREPARATION

The edges for a double seam are prepared in the same manner as a single seam. The edges for circular jobs are prepared on the burring or turning machine. The edges for square or rectangular jobs are prepared in the standard hand brake or bar holder.

BENDING CAUTIONS

When bending the edge over to make a double seam, a mallet should be used because a hammer stretches the metal. A double seam should be turned over and dressed to its proper shape with as little working of the metal as possible. If the metal is struck too often, the coating may flake or the edge may crack open, figure 28-6. These faults develop very quickly if the metal is hard or a poor grade.

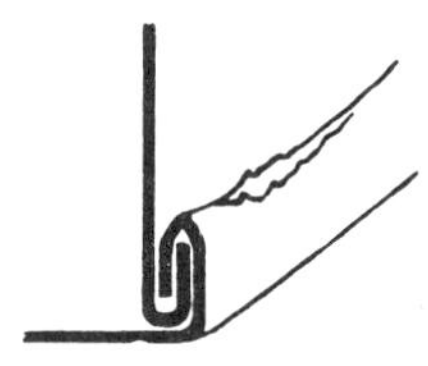

Figure 28-6 Seam cracked open.

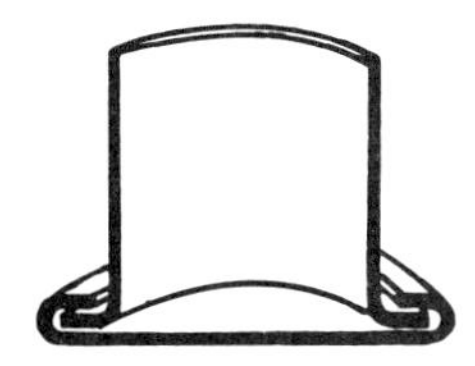

Figure 28-7 Single seam.

HOW TO DOUBLE SEAM A BOTTOM ON A CYLINDER

1. Burr the edge of the body and the bottom.
2. Make the single seam, figure 28-7, as in the preceding unit.
3. Place the job on the double-seaming stake with the cylinder over the stake.

If the curvature on the double-seaming stake is too large or too small, use a stake of a suitable curvature.

4. Hold the job firmly against the end of the stake with the left hand and bend the single seam to approximately a 45-degree angle by striking inward blows with a mallet, figure 28-8.

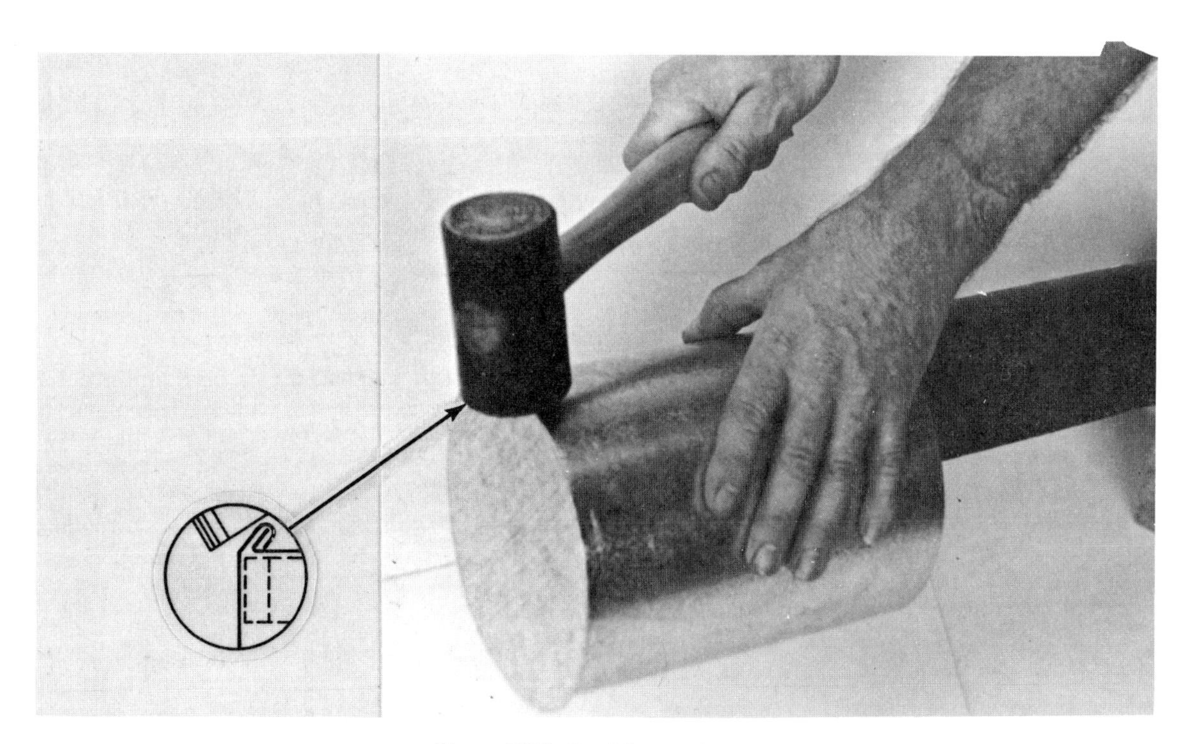

Figure 28-8 Double seaming.

With each blow of the mallet, gradually turn the job, keeping the cylinder and the bottom of the job firmly against the stake. If the job is not held in this position, the seam will crush.

5. Finish bending the edge with the mallet until the seam is flat.
6. Remove the job from the double-seaming stake and place it on the square head stake.

Figure 28-9 Squaring the edge.

7. Tap the bottom with the mallet to square the edge and straighten the seam, figure 28-9.
8. Replace the cylinder on the double-seaming stake and flatten the seam if necessary.

When double seaming a large job, the seam can be flattened by placing it on the flat surface of a stake, plate, or rail and tapping the inside of the job with the face of a setting hammer, figure 28-10.

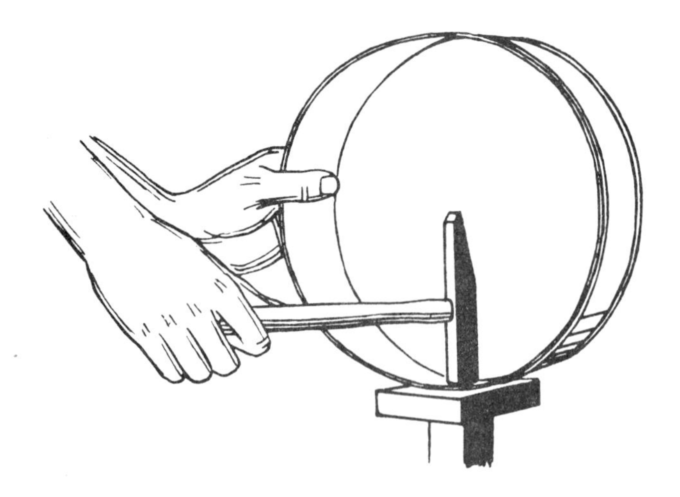

Figure 28-10 Finishing the seam.

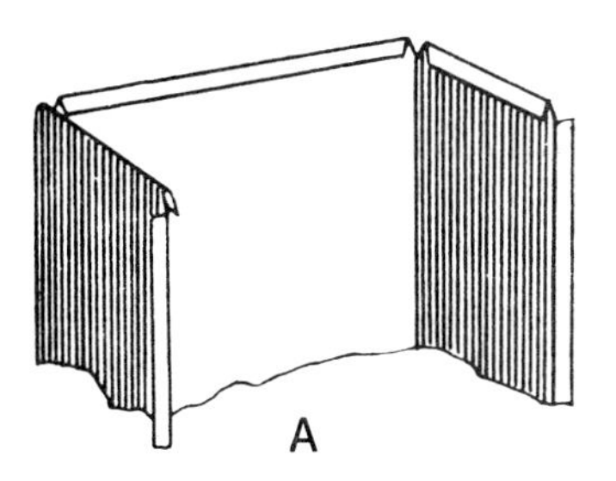

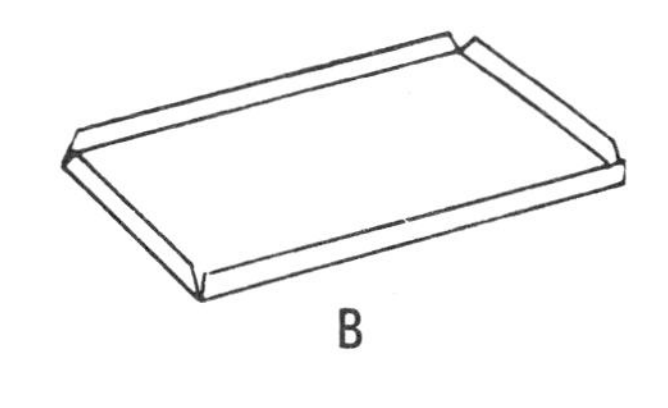

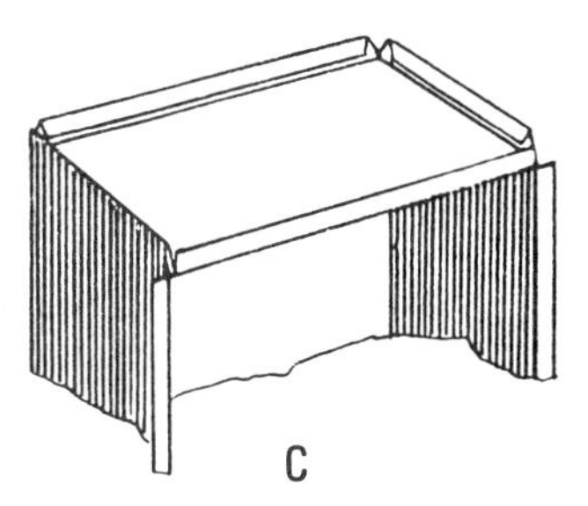

Figure 28-11 Assembling the box.

HOW TO DOUBLE SEAM CORNERS ON SQUARE OR RECTANGULAR JOBS

1. Brake the edges of the job in the standard hand brake, figure 28-11A.
2. Brake the edges of the end piece in the standard hand brake, figure 28-11B.
3. Put the end piece in the job, figure 28-11C.
4. Place the end of the job over the square stake. Hold the job firmly on the stake with the left hand. Starting at the top of the job, bend the seam down flat with a mallet, figure 28-12.

If the square stake is too large, use the hollow mandrel, beakhorn, or other suitable stake.

5. Repeat step 4 on the opposite side of the same end.

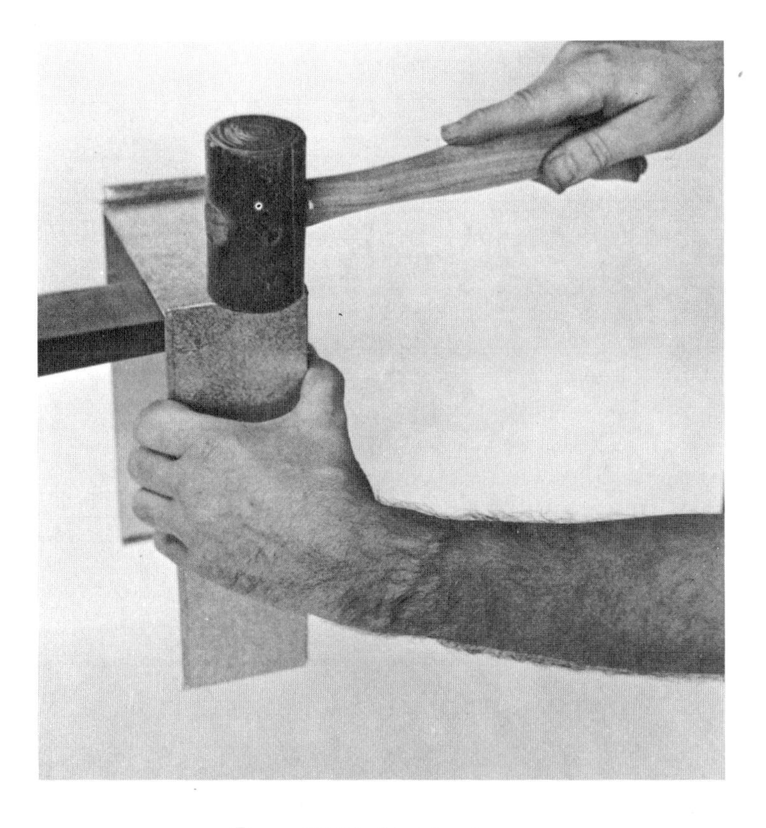

Figure 28-12 Starting the seam.

Figure 28-13 Squaring the edge.

6. Remove the job from the stake and place it in the position shown in figure 28-13. Tap the side with a mallet to square the edge and straighten the seam. Repeat on the opposite side of the same end.
7. Hold the job on the stake as shown in figure 28-14. Strike the seam from the inside with a setting hammer to tighten and smooth the seam. Repeat on the opposite side of the same end.
8. Repeat all operations on the bottom seam of the same end.
9. Repeat all operations on the other end.

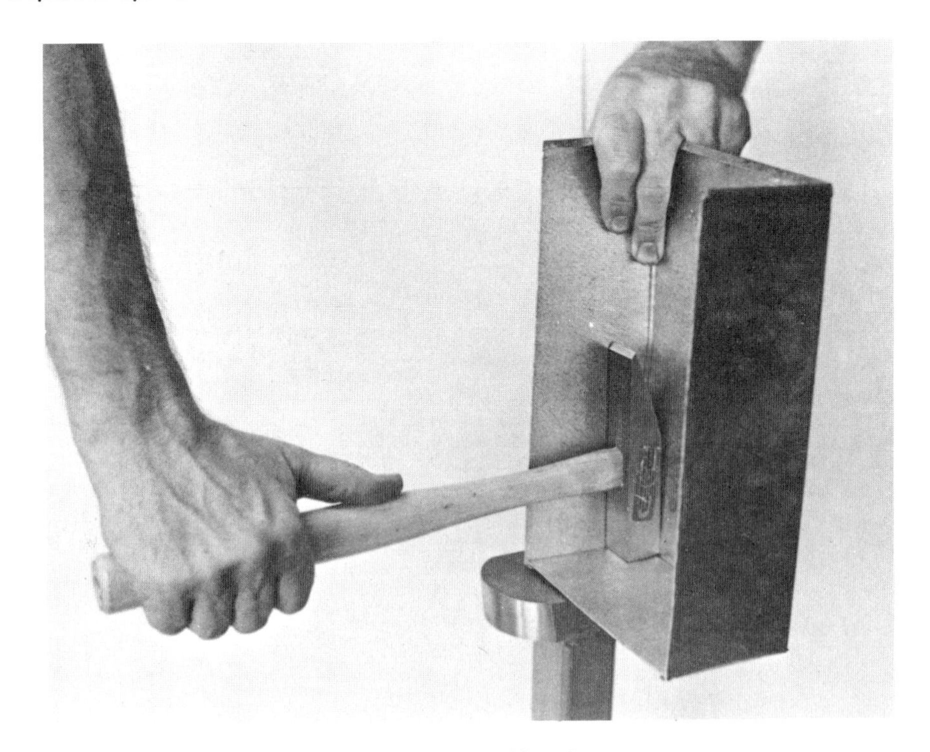

Figure 28-14 Smoothing the seam.

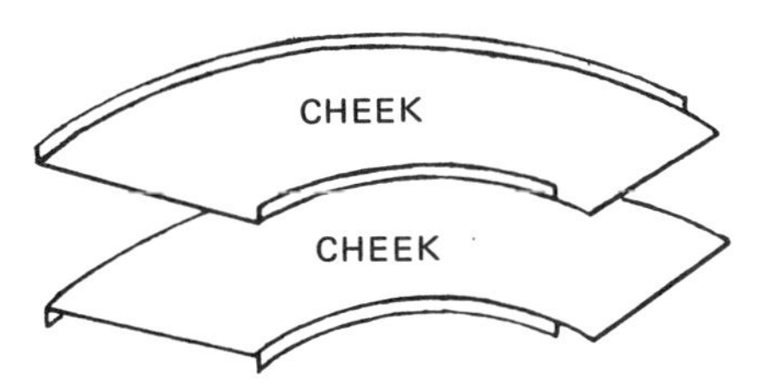

Figure 28-15 Cheeks.

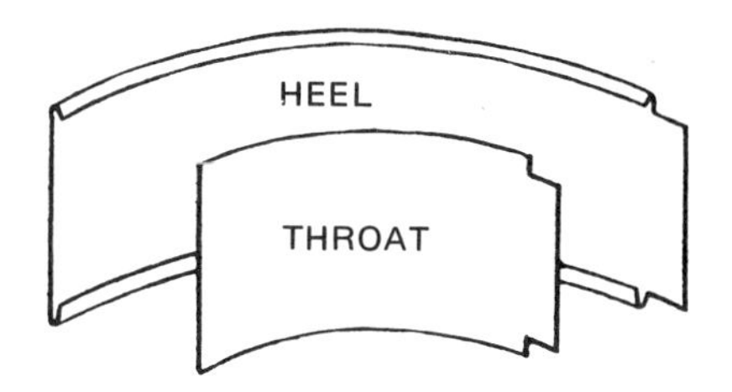

Figure 28-16 Heel and throat.

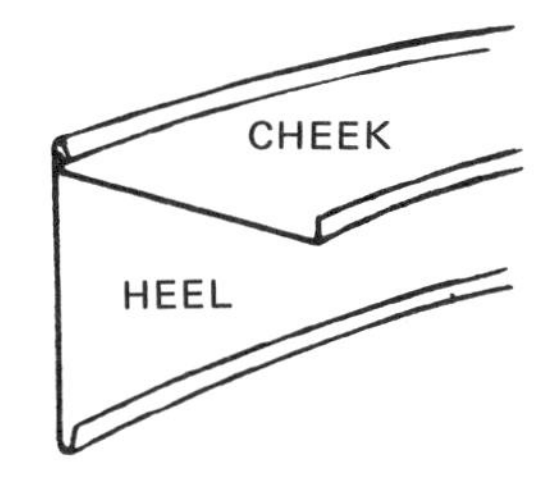

Figure 28-17 Cheek and heel.

HOW TO DOUBLE SEAM THE HEEL AND THROAT ON AN ELBOW

1. Flange the edges on the cheeks, figure 28-15.
2. Brake the edges of the heel and throat, figure 28-16, in the standard hand brake and roll them in the slip roll forming machine with strips of scrap metal inserted in the formed edges.
3. Place one of the cheeks in the heel, figure 28-17.
4. Place the job on the head of the double-seaming stake. Hold the job firmly against the end of the stake as shown in figure 28-18, and bend the seam over in several places with a mallet.

If the curvature on this head is too large or too small, use a stake of a suitable curvature.

5. Finish bending the seam until the seam is parallel to the cheek.

CAUTION

With each stroke of the mallet, gradually move the job along the stake, keeping the heel and the cheek firmly against the stake.

Figure 28-18 Tacking the seam.

Figure 28-19
Straightening the seam.

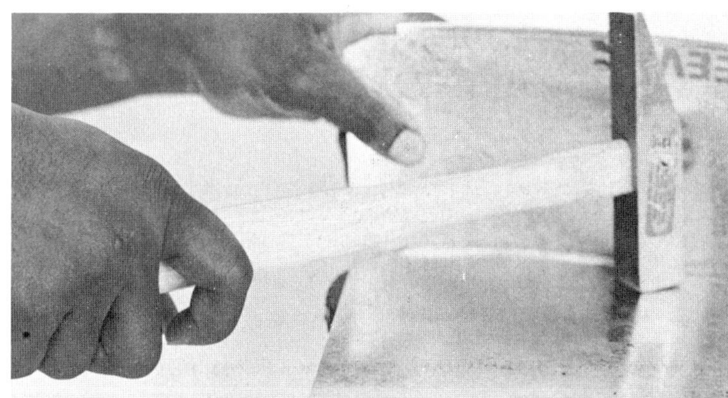

Figure 28-20
Flattening the seam.

Figure 28-21
Tacking the throat.

6. Back the seam with a dolly and tap it with a mallet as shown in figure 28-19. This operation squares the edge and straightens the seam.
7. Hold the job on the stake as shown in figure 28-20 and strike the seam from the inside with a setting hammer to flatten and smooth it.
8. Repeat steps 3 through 7 on the other end of the heel.
9. Bend the throat over in several places with a mallet, figure 28-21.
10. Repeat steps 3 through 7 on the throat.
11. Repeat steps 3 through 7 on the other end of the throat.

SUMMARY REVIEW

A. Place the answers to the following questions in the column to the right.

1. List four jobs that require a double seam. 1. ____________
____
____
____

2. List four requirements for an adequate double seam. 2. ____________
____
____
____

3. What is the general purpose of a double seam? 3. ____________

4. How is a double seam made watertight? 4. ____________

5. If a double seam is bent with a hammer, what two defects may occur? 5. ____________
____

6. List two tools which would be used to back up the double seam when hammering it over. 6. ____________
____

7. Give two reasons why a narrow double seam is used on circular jobs. 7. ____________
____

B. Insert the correct word in each of the following sentences.

1. A double seam is not waterproof but it is ____________ proof.
2. In hammering over a double seam, the metal ____________.
3. The process of bending over the double seam in several places is known as ____________ it.
4. The flat side of an elbow is known as the ____________.
5. The ____________ of an elbow is the wrapper around the outer portion.
6. The ____________ of a double seam is always put on the elbow wrapper.

C. Underline the correct word in each of the following sentences.

1. The double seam is always dressed down with a (setting hammer, rivet hammer, mallet) to minimize stretching.
2. The distribution allowance on the pocket of a double seam is (1/8, 1/16, 1/32) inch.
3. The flange allowance on a 1/4-inch double seam is (1/32, 1/8, 1/4, 9/32) inch.
4. The pocket allowance on a 1/4-inch double seam is (1/4, 1/8, 5/32, 1/2) inch.

UNIT 29 THE PITTSBURGH LOCK

OBJECTIVES

After studying this unit, the student should be able to

- List the uses and advantages of a Pittsburgh lock.
- Compute the allowances for a Pittsburgh lock.
- Fabricate a Pittsburgh lock on a fitting.

The Pittsburgh lock is a corner seam, figure 29-1. The edge of the pocket extends around the flanged piece to form the lock. This seam is sometimes called a *hobo lock* or *hammer lock* and is used extensively on heating and ventilating work.

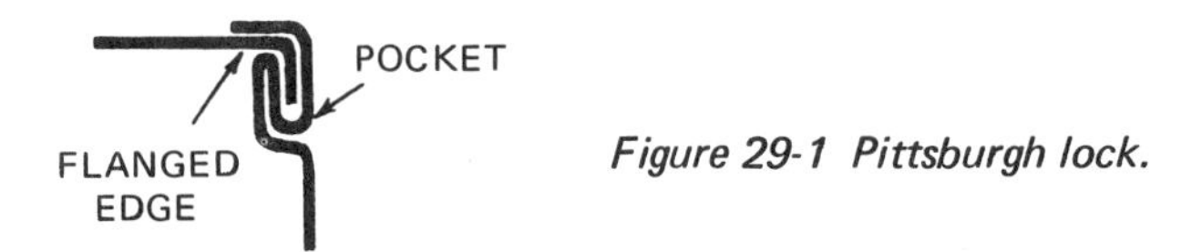

Figure 29-1 Pittsburgh lock.

USES

The Pittsburgh lock is used for pipe and fittings because the parts can be made in the shop and easily assembled on the job without the use of stakes of any kind. Since the unassembled parts occupy less space, the cost of shipping the material to the job is reduced. This seam is sometimes used instead of a double seam on many jobs, such as machine guards and boxes.

ALLOWANCES

The allowance for the pocket part of one side of a job is equal to two times the width of the pocket plus 3/16 inch, figure 29-2.

Figure 29-2 Pittsburgh seam allowances.

Example: Find the allowance for the pocket part of one side of a job with a Pittsburgh lock if the width of the pocket is to be 1/2 inch.

Solution:

$$W = \frac{1}{2}$$

$$A = 2W + \frac{3}{16}$$

$$A = \overset{1}{\cancel{2}} \times \frac{1}{\cancel{2}} + \frac{3}{16}$$

$$A = 1\frac{3}{16} \text{ inch, } \textit{Answer}$$

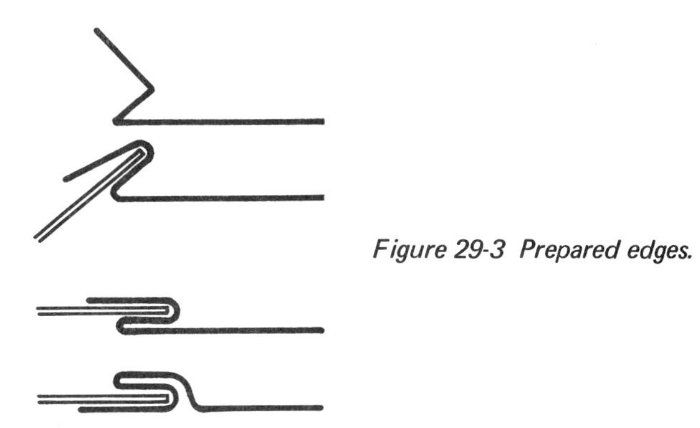

Figure 29-3 Prepared edges.

EDGE PREPARATION

The pocket for a Pittsburgh lock, figure 29-3, can be prepared in the standard hand brake, but it is usually done is a Pittsburgh lock former.

The flange is formed on the hand brake or bar folder. For curved work, it is formed on the easy edger. The width of the flange should always be less than the depth of the pocket to insure a tight seam.

HOW TO FINISH A PITTSBURGH LOCK

1. Set the flange into the pocket, and tap it with a hammer until it is firmly in place, figure 29-4.
2. Using a mallet, tack the seam by bending over the lap edge in several spots.
3. Finish the seam by bending down the entire lap edge with a mallet, figure 29-5.

Figure 29-4 Edges joined.

Figure 29-5 Finished lock.

HOW TO ASSEMBLE AN ELBOW WITH A PITTSBURGH LOCK

1. Brake the pockets of the heel and throat, figure 29-6.
2. Flange the edges of the cheeks, figure 29-7.
3. Place strips of scrap metal in the heel and throat and roll to the proper curvature, figure 29-8.

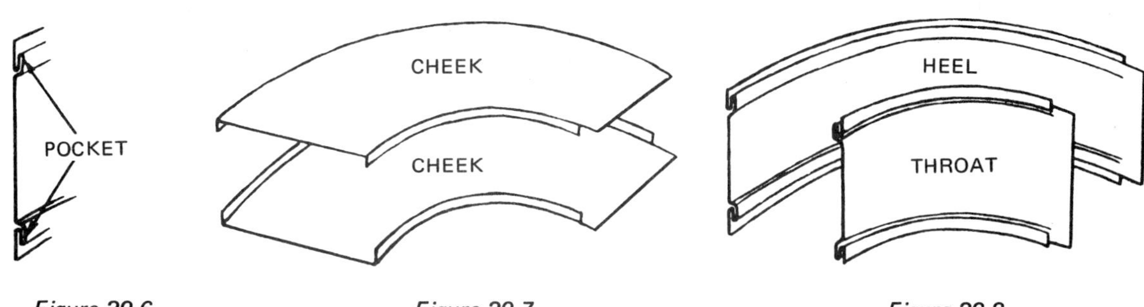

Figure 29-6

Figure 29-7

Figure 29-8

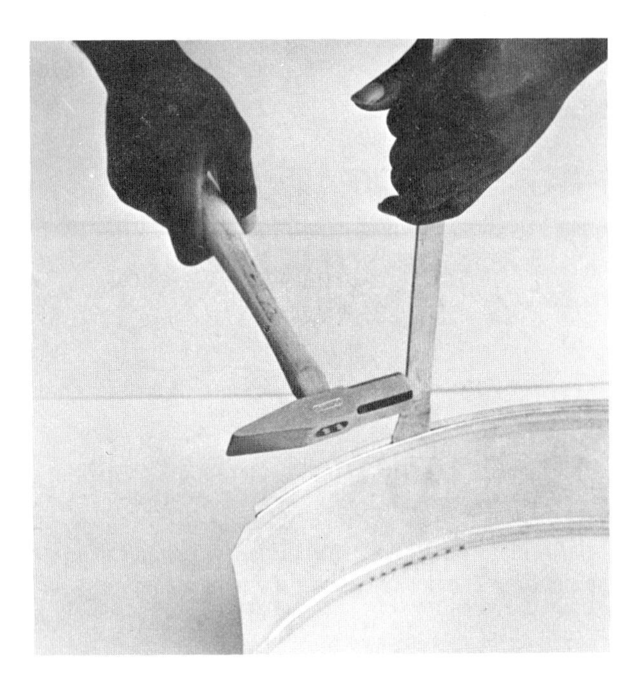

Figure 29-9 Opening the pocket.

4. Open the pockets with a pipe knife or seam opener and hammer, figure 29-9.

It is unnecessary to open the pocket with a pipe knife if a strip of scrap metal is inserted in the pocket before the rolling of the heel and throat.

5. Starting at one end, place the flanged edge into the pocket of the heel and bend over the projecting edge with a mallet.
6. Continue working the flange into the pocket with a rivet set and a mallet, and bend the edge over in several places with a mallet, figure 29-10.
7. Repeat steps 5 and 6 on the throat.

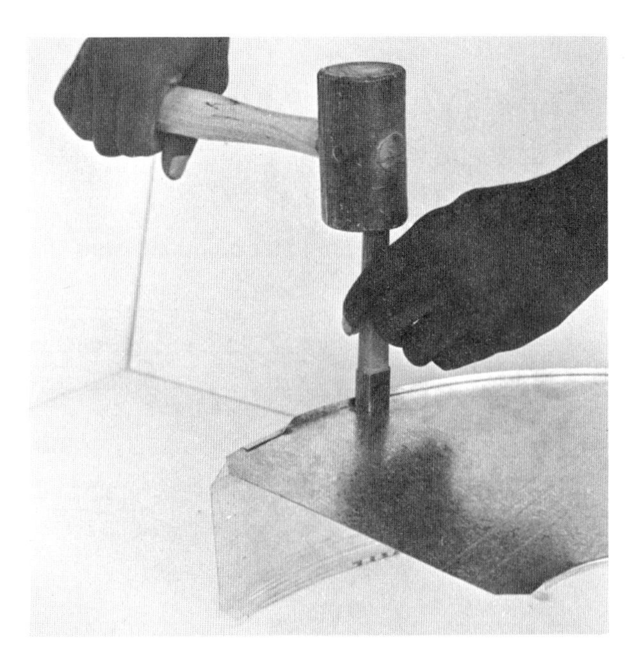

Figure 29-10 Working the flange into the pocket.

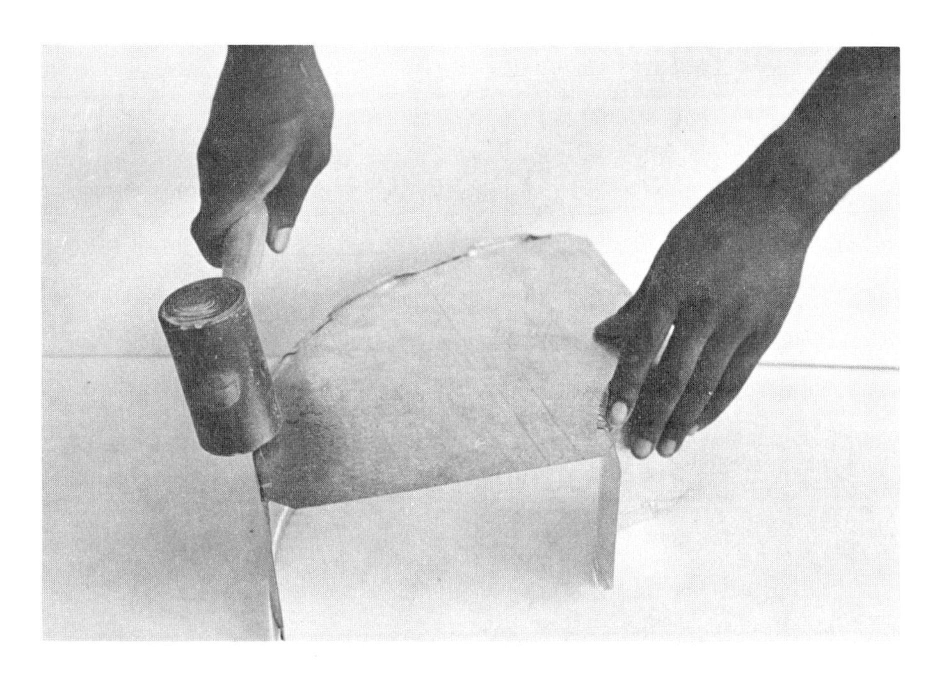

Figure 29-11 Bending the edge.

8. Finish bending the edge with a mallet on both the heel and throat until the edge is flat and smooth, figure 29-11.
9. Turn the job over with the finished edge resting on the bench or on the floor if the job is large.
10. Repeat steps 5 through 8 on the other cheek.

SUMMARY REVIEW

A. Place the answers to the following question in the column to the right.

1. List four jobs suitable for a Pittsburgh lock. 1. ______________________

   ____

   ____

   ____

2. List two advantages of using a Pittsburgh lock on heating and air conditioning jobs. 2. ______________________

   ____

3. Briefly state the steps in finishing a Pittsburgh lock. 3. ______________________

   ____

   ____

4. Name the two parts of the Pittsburgh lock. 4. ______________________

   ____

5. What is the Pittsburgh lock sometimes called in the field? 5. ______________________

B. Insert the correct word in each of the following sentences.

1. The Pittsburgh lock is not a ______________ seam.
2. The pocket of a Pittsburgh lock can be opened with a ______________, ______________ or ______________ and a hammer.
3. The Pittsburgh lock is a ______________ seam.
4. The pocket of a Pittsburgh lock can be made on a ______________ or a Pittsburgh lock former.

C. Underline the correct word in each of the following sentences.

1. The pocket allowance for a 5/16-inch Pittsburgh lock is (5/8, 15/16, 13/16) inch.
2. The flange allowance for a 5/16-inch Pittsburgh lock is (1/4, 5/16, 3/8, 9/32) inch.
3. A Pittsburgh lock cannot be used on (round pipe, square pipe, rectangular elbow).
4. The lap edge of a Pittsburgh lock is bent over with a (rivet hammer, rivet set, mallet).

UNIT 30 *GROOVED SEAMS*

OBJECTIVES

After studying this unit, the student should be able to

- List the uses for a grooved seam.
- Compute the allowances for a grooved seam.
- Make a grooved seam on pipe.

Grooved seams consist of two folded edges hooked together and offset. This type of seam is also called a *grooved lock* because it does not come apart if properly made. The grooved seam may be made by hand or machine. When the production of many pieces is required, the machine method is used.

USES AND TYPES

The grooved seam is used when making round pipes, square pipe, and containers. Pieces of metal may also be spliced with a grooved seam. Grooved seams are made by hooking together, offsetting, and flattening two folded edges, figure 30-1. The seams can be offset to the inside or outside of the job as illustrated in figure 30-1. The folds for both styles are the same. The outside seam is used on pipe and tank work in which the inside surface is to be flat. The inside seam is used in places where it is desirable to have the outside surface flat, as in hoods and panels.

The folded edges for a grooved seam should be of uniform width and folded scantly. The machines used for forming these edges are the bar folder for small work and the hand brake for large work.

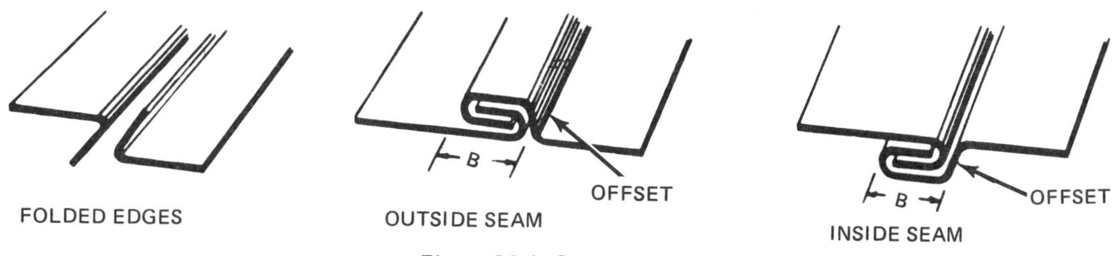

Figure 30-1 Grooved seams.

ALLOWANCES

Grooved seams are seldom formed on material heavier than 20 gage. For 24-gage stock and lighter, material three times the width W of the grooved lock is added to the pattern for the lock; one half of this allowance is added to each side of the pattern.

Formula: $A = 3W$.

For 23-gage stock and heavier, three times the width W plus four times the thickness T of the material, is added to the pattern; one half of this total allowance A is added to each side.

Formula: $A = 3W + 4T$.

HOW TO FINISH AN OUTSIDE GROOVED SEAM

Before finishing a grooved seam, check the edges for straightness, proper allowance, and tightness of fit.

1. Fold the edges to the proper width and form the job.
2. If the work is cylindrical, place the job on the hollow or solid mandrel stake. If the work is flat, place the job on a flat bar or plate.

If the curvature of these stakes is too large, use a stake of smaller curvature.

3. Hook the edges together, figure 30-2.

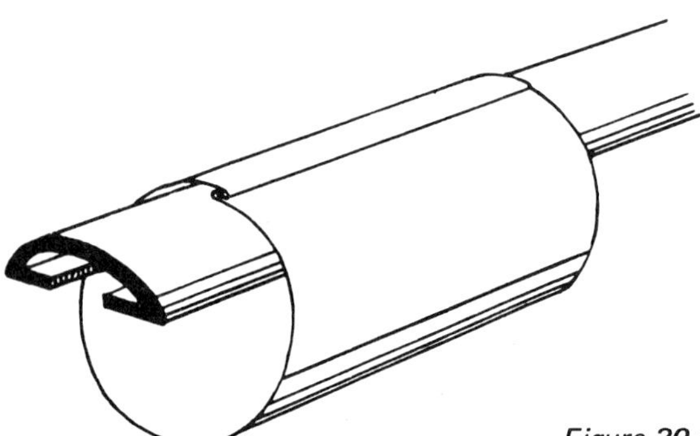

Figure 30-2 Job on stake.

4. Flatten the seam slightly with a mallet.
5. Place the hand groover over one end of the seam and strike it with a hammer.

The width of the groove in the hand groover must be slightly larger (about 1/16 inch) than the width of the seam to prevent cutting the material along the seam.

6. Groove the other end in the same manner, figure 30-3.

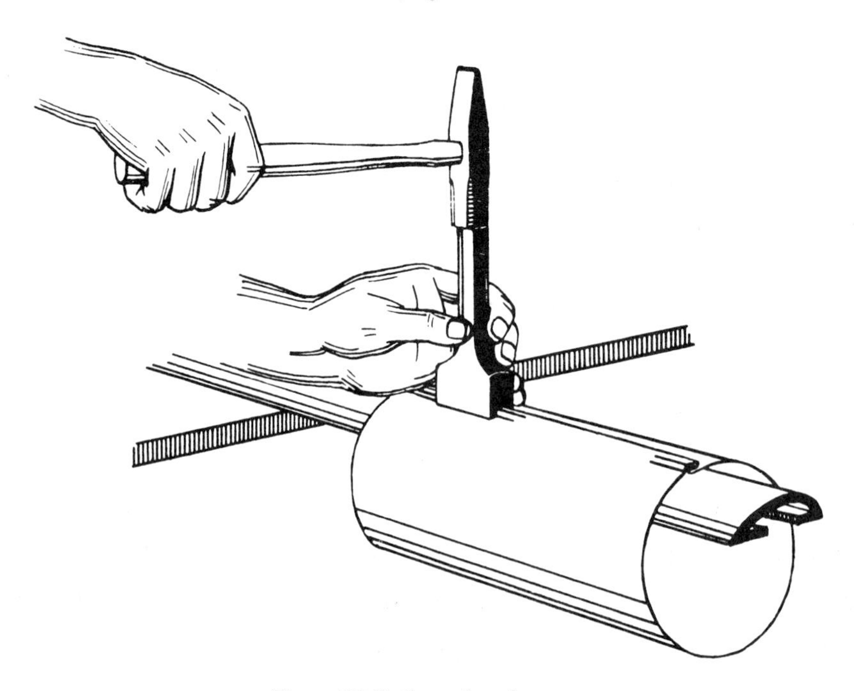

Figure 30-3 Grooving the seam.

7. Groove the entire seam by striking the hand groover with the hammer, while moving the groover along the seam.
8. Finish the seam by flattening it down closely with a mallet.
9. When the small end of a pipe (furnace and smoke pipe) is to be crimped, the seam is secured in one of the following ways:
   a. Prick punching at each end of the finished seam, figure 39-4A.
   b. Slitting the small or crimped end and bending the small piece of material back over the seam, figure 30-4B.
   c. Placing a small rivet in the seam at the small end, figure 30-4C.

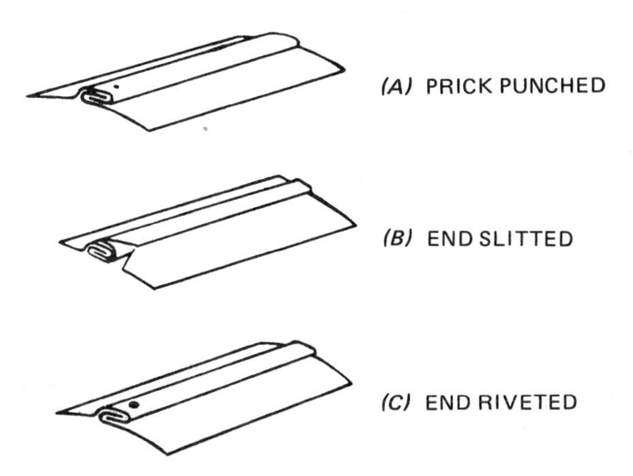

Figure 30-4 Locking the seam.

HOW TO FINISH AN INSIDE GROOVED SEAM

Before finishing a grooved seam, check the edges for straightness, proper allowances, and tightness of fit.

1. Fold the edges to the proper width and form the job.
2. Select the proper rail and fasten it to the bench.

The width of the groove in the grooving rail must be slightly larger (about 1/16 inch) than the width of the seam to prevent cutting the material along the seam.

3. Hook the edges together and place the job on the grooving rail.
4. Hold the seam in the groove on the grooving rail.
5. Strike near one end of the seam with a mallet, figure 30-5.
6. Repeat steps 4 and 5 on the opposite end of the seam.
7. Groove the entire seam by striking the metal with a mallet to force it into the groove on the rail.
8. Remove the job from the groove and place it on the flat part of the grooving rail.
9. Close the seam by striking it with a mallet, figure 30-6.

Figure 30-5
Grooving on a rail.

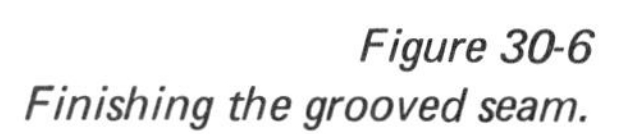

Figure 30-6
Finishing the grooved seam.

SUMMARY REVIEW

A. Place the answers to the following questions in the column to the right.

1. List three general uses for a grooved seam. 1. ______________________

   ____

   ____

2. List two jobs in which an outside grooved seam is used. 2. ______________________

   ____

3. List two jobs in which an inside grooved seam is used. 3. ______________________

   ____

4. Briefly state the steps in making a grooved seam. 4. ______________________

____

____

____

5. List three methods used to lock the grooved seam on a crimped stove pipe end. 5. ______________________

____

____

B. Insert the correct word in the following:

1. The inside grooved seam is used in places where the outside surface is to be kept ____________.
2. Edges are folded on the ______________________ for small work.
3. The ____________ of the metal is disregarded when figuring the groove seam allowance for metals 24-gage or lighter.
4. Grooved seams are seldom formed on materials heavier than ____________.

C. Underline the correct word in each of the following sentences.

1. The total allowance for a 5/16-inch grooved seam on 26-gage galvanized iron, .0179 inch thick, is (5/16, 15/16, 11/64) inch.
2. The total allowance for a 1/4-inch grooved seam on 22 gage iron, .0299 inch thick, is (3/4, 7/8, 1) inch.
3. An inside grooved seam on a found pipe is finished with a (hand groover, grooving rail, rivet set).

UNIT 31 *THE PLAIN DOVETAILED SEAM*

OBJECTIVES

After studying this unit, the student should be able to

- List the types of dovetail seams used.
- List the steps to follow in making a plain dovetail seam.
- Make a dovetail seam on a collar and flange.

Dovetailing is an easy and convenient method of joining collars and fittings to another part of the job without the use of solder, bolts, or rivets. There are several types of dovetail seams, such as the plain dovetail, the beaded dovetail, and the flanged dovetail. The plain dovetail will be the only one considered in this unit because it can be made in the field without a special tool or machine.

The plain dovetail seam is made by slitting or notching the end of a fitting and then bending out every other piece of metal, figure 31-1. The bent edges act as stops, and the remaining pieces are then bent over the other part of the job to act as locks, figure 31-2.

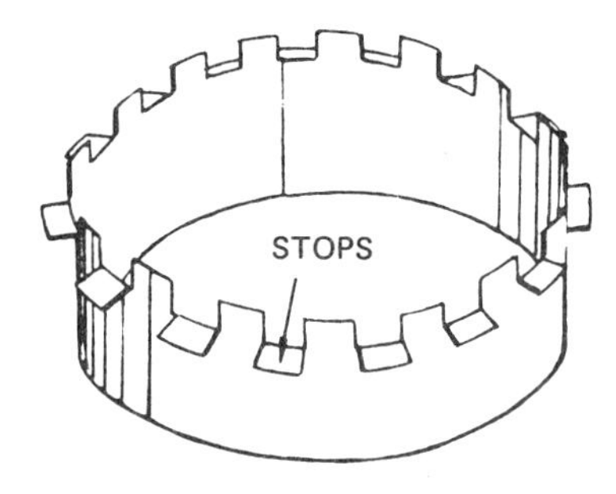

Figure 31-1 Dovetails.

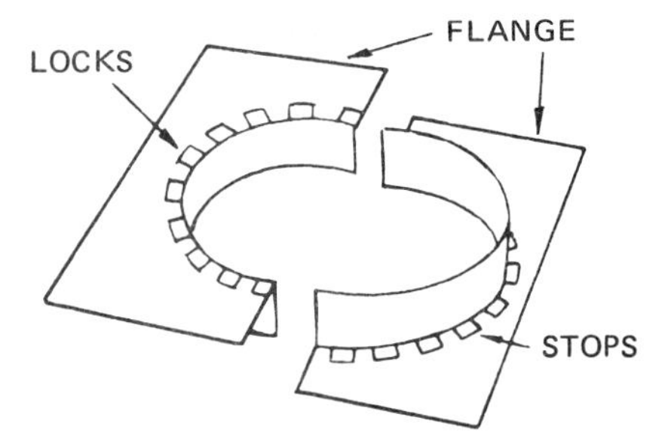

Figure 31-2 Finished dovetail.

Dovetailing is used to join a collar or pipe to another piece of pipe, to a flat surface (a flange), or to a curved surface. If the job is to be watertight, the seam is then soldered.

The width of the dovetail seam (depth for notching) can vary from 1/4 to 1/2 inch. The distance between the slits can vary from 1/4 to 1 inch, depending upon the size of the job. The slits are closer together on small jobs.

HOW TO MAKE A PLAIN DOVETAIL SEAM

1. Mark the width of the dovetail with a marking gage as shown in figure 31-3.
2. Notch the edge of the line at intervals with the snips. The distance between the slits can vary from 1/4 to 1 inch, depending upon the size of the job.
3. Bend the seam stop out at right angles with pliers as shown in figure 31-4, or hold the seam against the square stake and bend the seam stop with the peen end of the riveting hammer, figure 31-5.
4. Complete the rest of the stops by bending out alternate pieces with the pliers or the riveting hammer as in step 3.
5. Set the collar or pipe in the opening of the flange with the stops against the flange. To make a dovetail connection watertight, solder it on the outside, allowing the solder to flow into the dovetail. To make a dovetail seam at an angle, proceed in the same manner as for a plain dovetail seam.

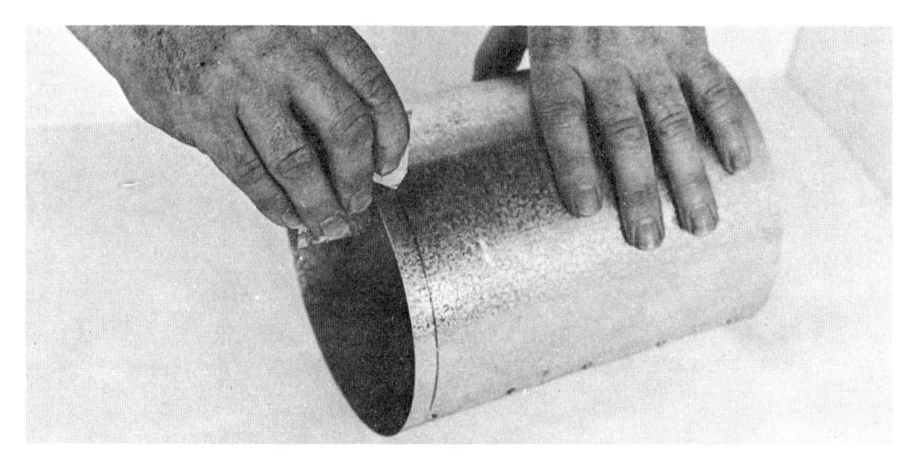

Figure 31-3 Marking with a gage.

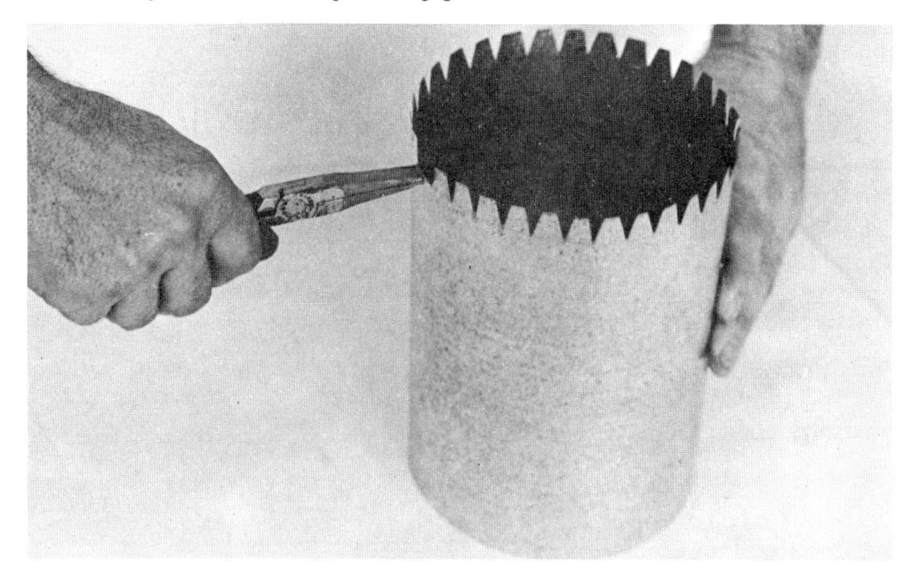

Figure 31-4 Bending seam stops with a pliers.

Figure 31-5 Bending stops with a hammer.

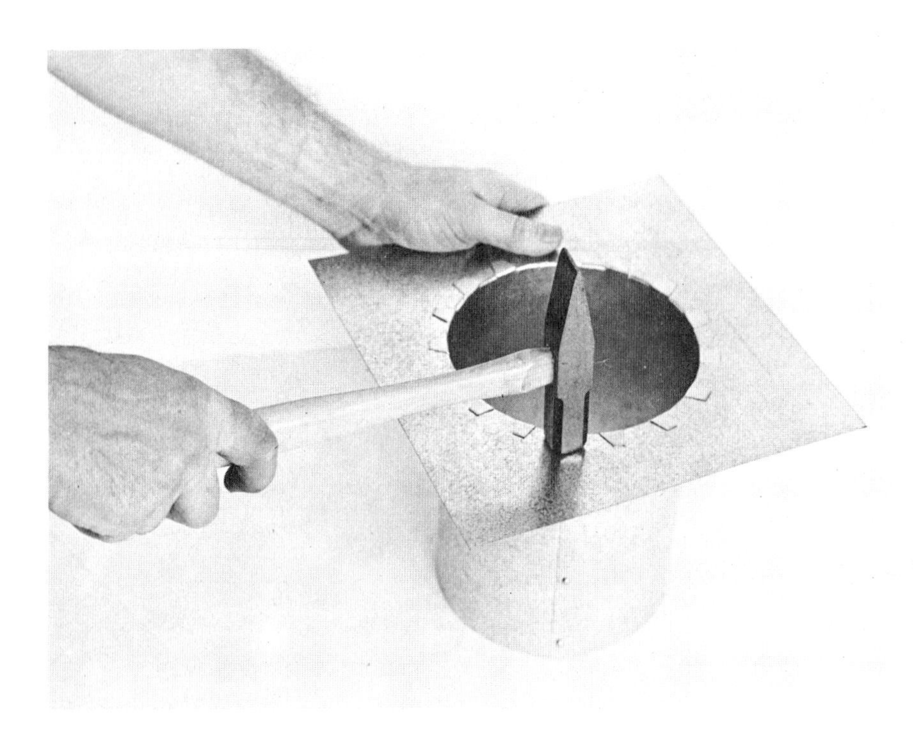

Figure 31-6 Bending the locks.

Figure 31-7 Flattening the stops.

6. Bend the upright edges over the flange to a 45-degree angle with the hammer, figure 31-6.
7. Bend the edges flat with the hammer to form the locks.
8. Place the flange on a flat surface and flatten the stops with the hammer, figure 31-7.

SUMMARY REVIEW

A. Place the answers to the following questions in the column to the right.

1. List the three types of dovetail seams used. 1. ____________________
____
____

2. Briefly state five steps to follow in making a dovetail seam. 2. ____________________
____
____
____
____

3. List the two parts of a dovetail seam. 3. ____________________
____

B. Insert the correct word in each of the following sentences.

1. Dovetailing is used to join ______________ and ______________ to another part of the job.
2. If a dovetail seam is to be watertight, it must be ______________ .
3. The following tools would be used in making a plain dovetail seam: ____________________
____.
4. The depth for the notching of a plain dovetail can vary from ______________ to ______________ inch.

C. Underline the correct word in each of the following sentences.

1. Notches are cut with (a hacksaw, side cutters, snips) for a plain dovetail seam.
2. A dovetail seam is used to join (sheets together, collars to flanges).

UNIT 32 *STANDING SEAMS*

OBJECTIVES

After studying this unit, the student should be able to

- List the uses for the standing seam.
- Compute the allowances for a standing seam.
- Make a standing seam properly.

The standing seam is a flat surface seam, figure 32-1. The flange is formed at 90 degrees to the flat surface, and the pocket slips over it. It can be secured with rivets, buttons, or bolts. The seam is not considered water-tight, but if properly used, it sheds water without leaking. To insure water tightness, solder or mastic is applied.

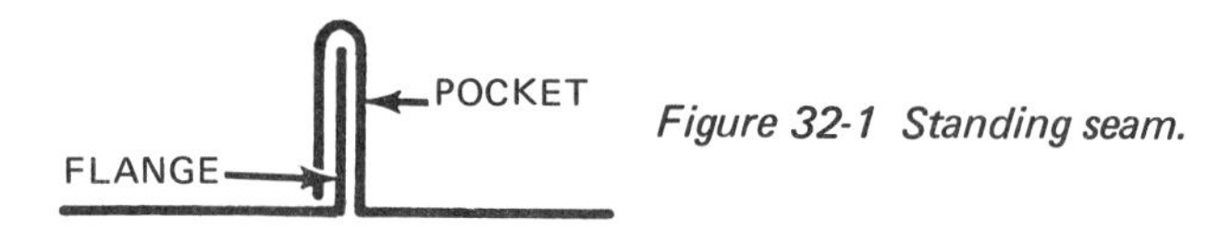

Figure 32-1 Standing seam.

USES

The standing seam is used as a stiffener in joining pieces of metal over large, flat areas. In large duct construction it is used with the seams on the inside to insure rigidity. The standing seam is installed easily on the pitched roofs of canopies, awnings, and weather sheds. It protects well and is strong. A special double-seamed standing seam is used on large pitched roofs.

ALLOWANCES

The total seam allowance is equal to three times the seam width (A = 3W). The pocket allowance equals 2*W*, and the flange allowance equals *W*.

In laying out the standing seam, figure 32-2, the pocket allowance is placed on one side and the flange allowance on the other. However, the first bend line on the pocket is made at *W* minus 1/8 inch to allow for metal thickness and to insure a tight seam.

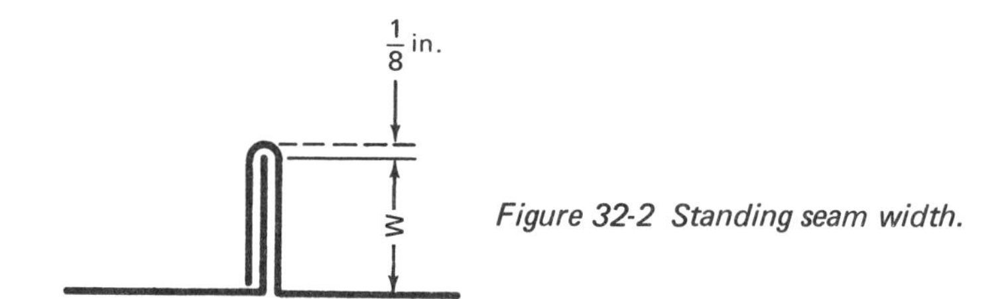

Figure 32-2 Standing seam width.

EDGE PREPARATION

After laying out the desired size of the standing seam, form the pocket and flange on the hand brake. A bar folder can be used for seams of small length and width. A piece of stock should be kept in the pocket while forming to prevent complete closure.

HOW TO FINISH A STANDING SEAM

1. Place the pocket over the flange and tap it down until firmly in place.
2. Secure with rivets, buttons, or bolts through the seam.

SUMMARY REVIEW

A. Place the answers to the following questions in the column to the right.

1. List two general purposes of a standing seam. 1. ______________________
____

2. List three jobs that require the standing seam in fabrication. 2. ______________________
____
____

3. List three reasons why a standard seam is used. 3. ______________________
____
____

B. Insert the correct word in each of the following sentences.

1. A standing seam is used on ______________ roofs.
2. A standing seam is not considered to be ______________.
3. In forming the pocket for a standing seam on the hand brake, a piece of ______________ should be kept in the pocket to prevent complete ______________.

C. Underline the correct word in the following:

1. The total allowance for a standard seam can be expressed as (W, 2W, 3W).
2. The first bend line in forming the standing seam pocket should be equal to (W, W plus 1/8 inch, W minus 1/8 inch).
3. A standing seam can be secured with (clamps, tongs, tape, rivets).
4. The allowance for a 1-1/4 inch standing seam on a canopy is (1-1/4, 2-1/2, 3-3/4) inch.

UNIT 33 *WIRE EDGES*

OBJECTIVES

After studying this unit, the student should be able to

- Identify and describe the types of wire jobs.
- Compute wire edge allowances.
- Make a wire edge on square and round jobs.

The edges of some jobs are folded around wire to strengthen and stiffen them and to eliminate sharp edges. Wired edges, figure 33-1, can be made by hand or machine. The hand process is used when a machine is not available, when the work is too large for a machine, and when the material is too heavy for the capacity of the machine. A sheet metal worker is often called upon to do hand wiring.

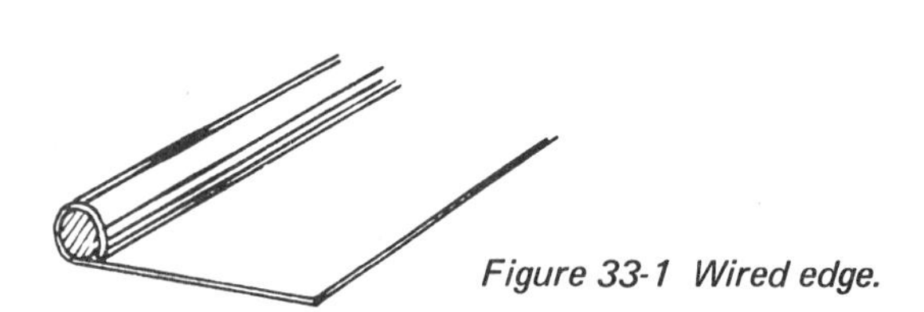

Figure 33-1 Wired edge.

Wire is used in sheet metal work to reinforce edges, support furnace pipes, and make handles and hinges. Usually, uncoated wire is used for jobs of black iron. Tinned or copper coated wire is used for tinware jobs, and galvanized wire is used for galvanized jobs.

The sheet metal worker uses an annealed steel wire which is strong and more suitable for sheet metal jobs than other types. Wire is obtainable in coils of various weights. It can be ordered from the mill according to gage number or diameter size. Round wire stock in straight lengths is called *rod* and can be obtained in lengths up to 20 feet. Rod is used on straight jobs, and wire is preferred for curved jobs. Because wire is less expensive than rod, it is used more often.

If wire is cut from a coil, wrap a piece of small wire or tape around the coil to prevent loosening and to keep the ends from sticking out. Protruding ends can cause serious injury.

EDGE PREPARATION

Where the wired edge crosses a seam, some of the overlapping metal must be cut away to eliminate a bulge at the seam. This is called *notching* and must be done carefully before the job is formed and wired to prevent leaving a bulge or hole at the seam.

All wired edges are not made at the same stage in the construction of the job. Some jobs are wired before they are formed, and others are wired after they are formed. Cylindrical jobs having straight sides, such as cans or tanks, and the small ends of tapering jobs are usually wired before the jobs are formed into shape. Boxes, machine guards, and the large ends of tapering jobs, such as pails, are wired after they are formed. The wired edge is always put on the outside of the job unless otherwise specified. This edge can be made on a turning machine, bar folder, standard hand brake, or by hand, figure 33-2.

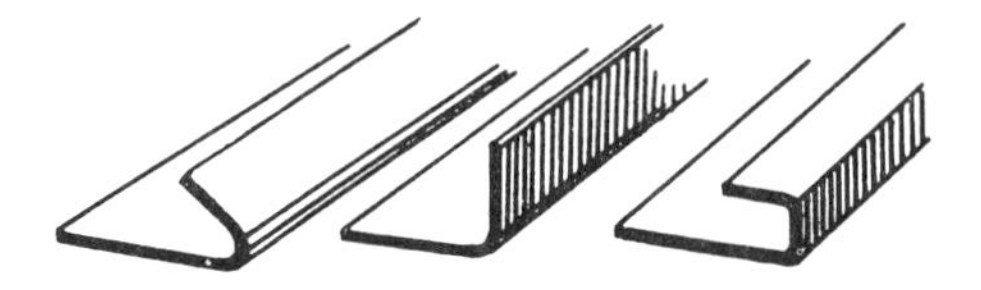

Figure 33-2 Prepared edges.

SHEET METAL ALLOWANCE

For 24-gage stock and lighter, material 2-1/2 times the diameter of the wire is added to the pattern for the wired edge.

Example: Find the allowance to be added to the pattern for 1/8-inch wire.

Solution: $A = \frac{1}{8} \times 2\frac{1}{2}$

$A = \frac{1}{8} \times \frac{5}{2} = \frac{5}{16}$ inch, *Answer*

WIRE ALLOWANCES

The length of the wire needed for a wired edge is equal to the *perimeter* (distance around) of the job with an extra allowance for the thickness of the wire. Formulas and examples are given below.

SQUARE JOBS

Rule: The length of the wire (approximately) is equal to 4 times one side of the job plus 2 times the diameter of the wire.

Formula: L.W. = 4S + 2d

L.W. = length of wire in inches
S = length of one side in inches
d = diameter of wire in inches

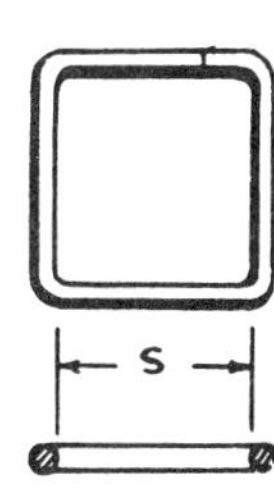

Figure 33-3
Square job.

Example: If S = 18 inches and $d = \frac{1}{4}$ inch, find L.W.

1. L.W. = 4S + 2d
2. L.W. = $(4 \times 18) + (2 \times \frac{1}{4})$
3. L.W. = $72 + \frac{1}{2} = 72\frac{1}{2}$ inches, *Answer*

RECTANGULAR JOBS

Rule: The length of the wire (approximately) is equal to 2 times the length plus 2 times the width of the job plus 2 times the diameter of the wire.

Formula: L.W. = 2L + 2W + 2d

L.W. = length of wire in inches
L = length of job in inches
W = width of job in inches
d = diameter of wire in inches

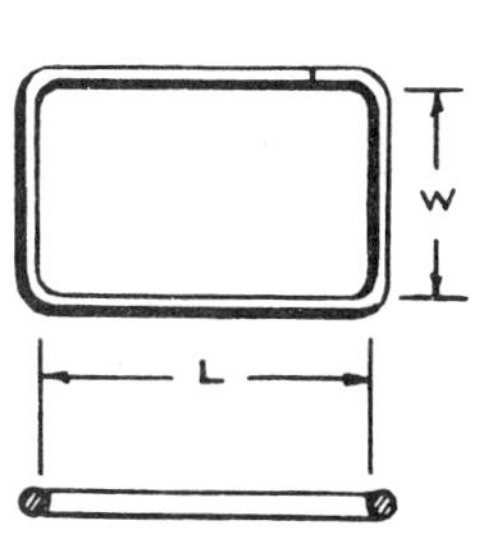

Figure 33-4
Rectangular job.

Example: If L = 10 inches, W = 5 inches, and $d = \frac{1}{8}$ inch, find L.W.

1. L.W. = 2L + 2W + 2d
2. L.W. = $(2 \times 10) + (2 \times 5) + (2 \times \frac{1}{8})$
3. L.W. = $20 + 10 + \frac{1}{4} = 30\frac{1}{4}$ inches, *Answer*

CIRCULAR JOBS

Rule: The length of the wire is equal to the sum of the diameter of the job and the diameter of the wire multiplied by π(pi) which equals 3.1416. If wiring is done before the job is formed, use πD.

Formula: L.W. = π(D + d)
π = 3.1416
D = diameter of job in inches
d = diameter of wire in inches

Example: If D = $19\frac{3}{4}$ inches and d = $\frac{1}{4}$ inches, find L.W.

1. L.W. = π(D + d)
2. L.W. = 3.1416 ($19\frac{3}{4} + \frac{1}{4}$)
3. L.W. = 3.1416 x 20
4. L.W. = 62.832 inches = $62\frac{27}{32}$ inches, *Answer*

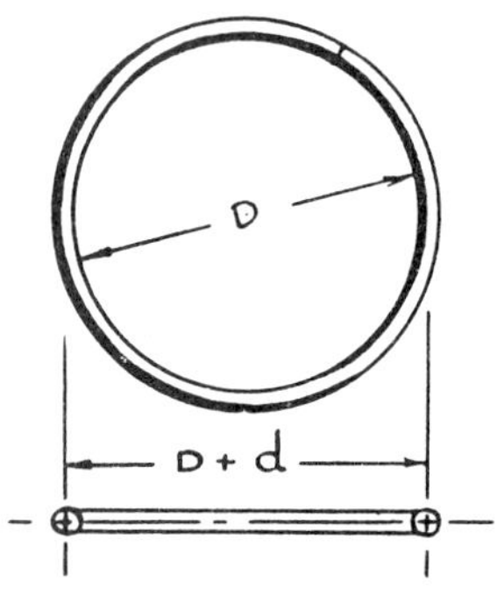

Figure 33-5
Circular job.

HOW TO WIRE A STRAIGHT EDGE BEFORE THE JOB IS FORMED

1. Make the edge of the job, figure 33-6, equal to 2-1/2 times the diameter of the wire.

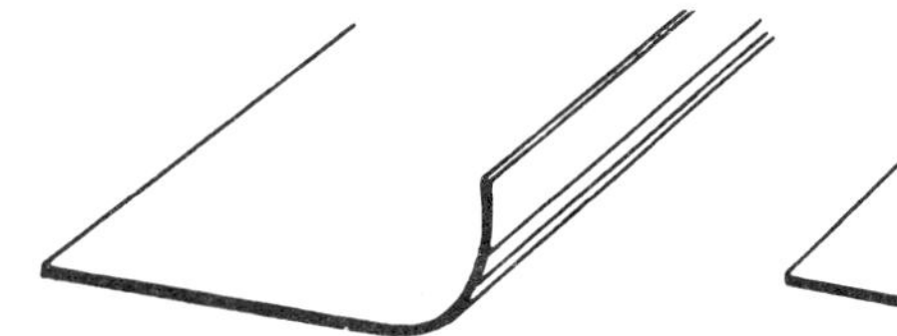
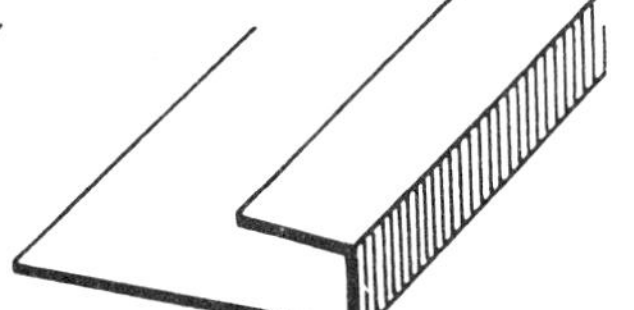
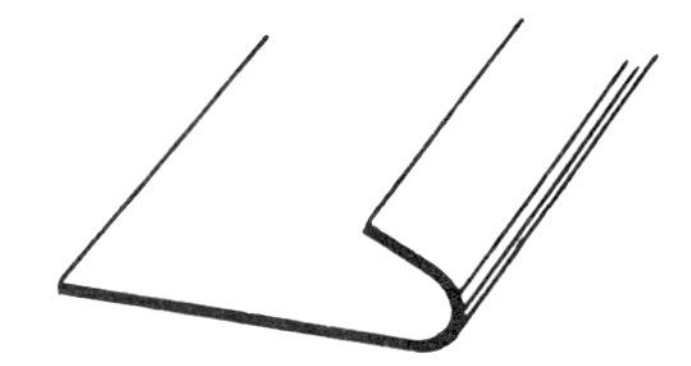

Figure 33-6 Types of edges.

2. Cut the wire to the correct length with a pair of pliers or wire nippers.

Wire should not be cut with snips or on the squaring shear because it nicks the cutting edges.

3. Lay the job on a flat plate or work bench. If the job has a grooved seam, fold the edges for the grooved seam before inserting the wire.
4. Straighten the wire and place it into the fold. For a cylinder, extend the wire 1 to 2 inches from the underfold end. This is done to strengthen the seam and reduce the possibility of an opening remaining after the job is formed.
5. Start the fold over the wire the entire length of the edge, using a mallet and pliers, figure 33-7.

Use pliers to prevent injury to your fingers.

Figure 33-7 Holding wire with pliers.

6. Bend the metal over the wire as far as possible with the mallet. Figure 33-8 shows the three steps necessary to make a wired edge.

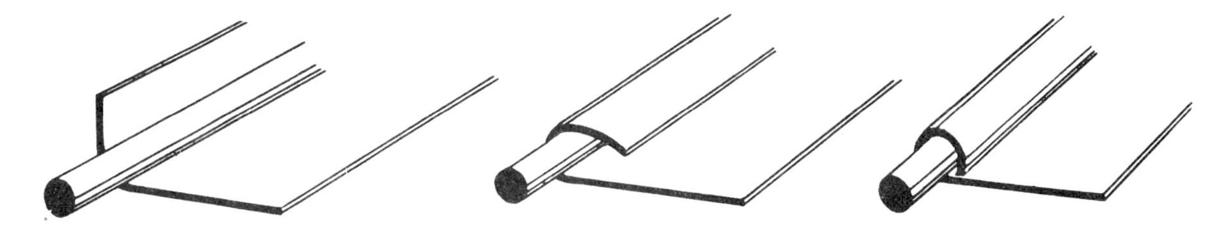

Figure 33-8 Steps in making a wired edge.

7. With the peen of the setting hammer, complete the wired edge by peening the metal down tightly over the wire, figure 33-9.

Be careful not to mar the job with the mallet or the setting hammer.
The work must be backed with a rail or a flat plate during the entire wiring operation

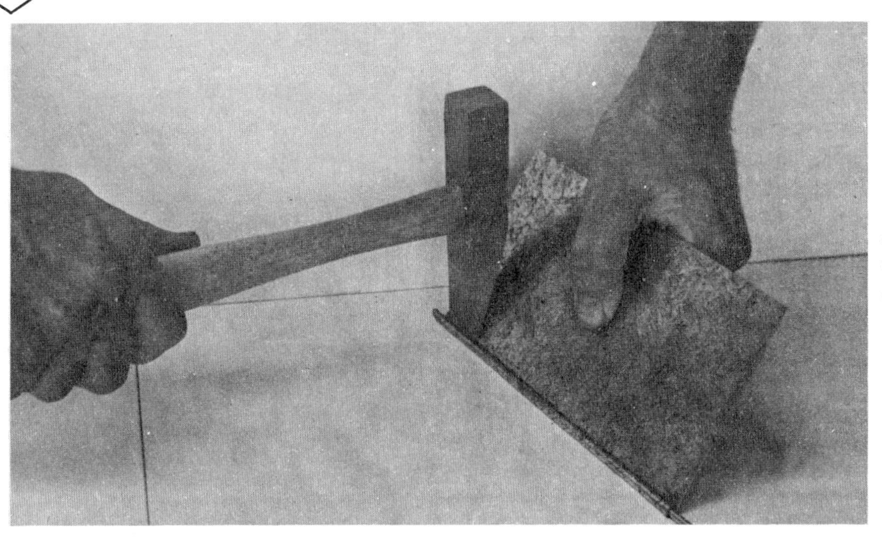

Figure 33-9 Completing a wired edge.

Figure 33-10 Turned edge.

HOW TO WIRE AN INSIDE OR OUTSIDE RADIUS ON A FLAT JOB

1. Turn the edge equal to 2-1/2 times the diameter of the wire, figure 33-10.
2. Measure the length of the edge to be wired with a flexible rule and cut the wire to this length with pliers or wire nippers.
3. Roll the wire into the required shape in the groove of the slip roll forming machine.
4. Lay the job on a flat plate or work bench and place the wire in the fold. If the job is to be rolled, keep the wire at least 1 to 2 inches from one end.
5. Start the fold over the wire the entire length of the edge, using a mallet and pliers, figure 33-11.

Use pliers to prevent injury to your fingers.

6. Bend the metal over the wire as far as possible with the hammer, figure 33-12.

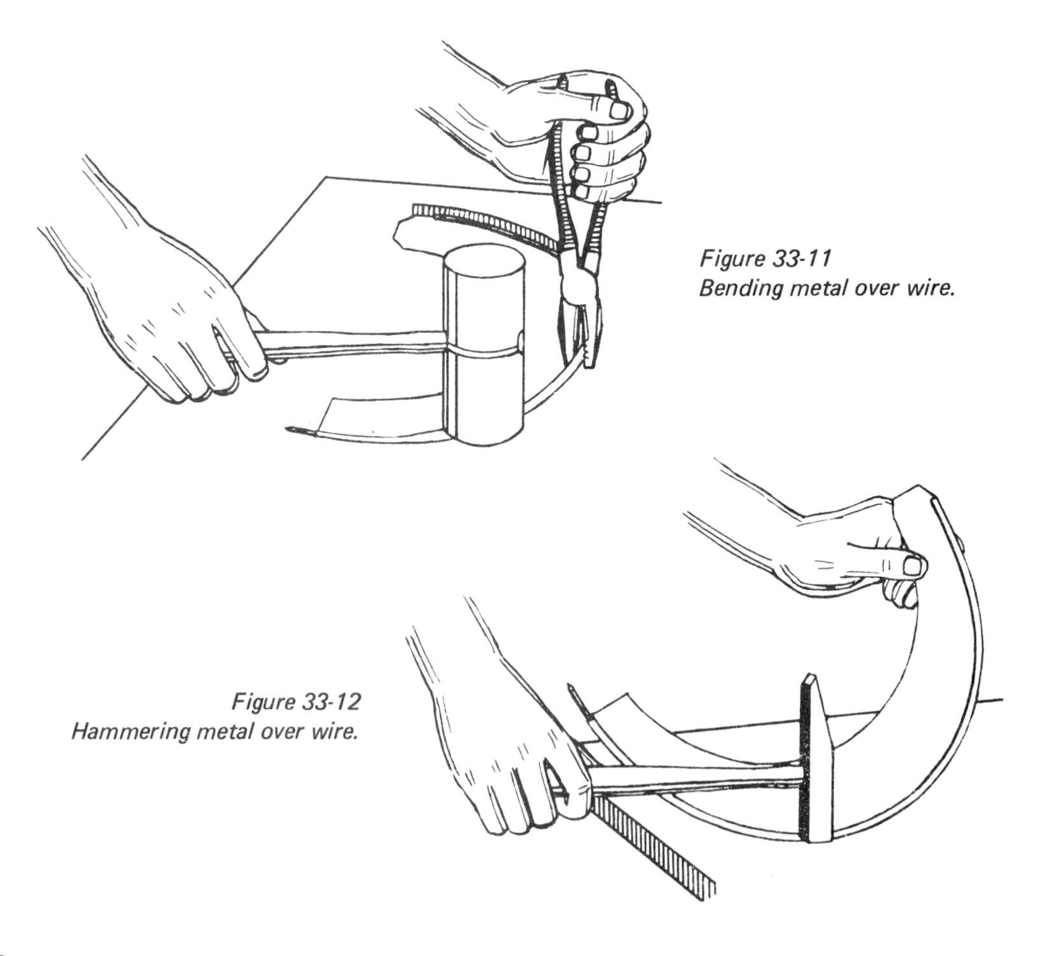

Figure 33-11
Bending metal over wire.

Figure 33-12
Hammering metal over wire.

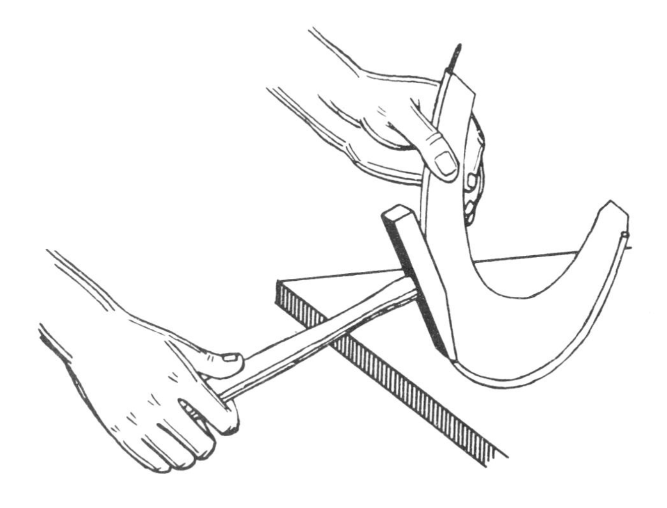

Figure 33-13 Finishing the edge.

7. With the peen of the setting hammer, complete the wired edge by peening the metal down tightly over the wire, figure 33-13.

Be careful not to mar the job with the hammer or mallet.

HOW TO WIRE A SQUARE OR RECTANGULAR JOB AFTER IT IS FORMED

1. Make the edge of the box equal to 2-1/2 times the wire diameter, figure 33-14.

Figure 33-14 Forming a box edge.

2. Cut the wire to the correct length with a pair of pliers or wire nippers.

Wire should not be cut with snips or on the squaring shear because it nicks the cutting edges.

3. Place the wire in the vise with 1 to 2 inches extending at one end. The ends of the wire should meet 1 to 2 inches from the corner of the job, figure 33-15. This strengthens the corner and improves the appearance.
4. Bend the end of the wire to a 90-degree angle with a setting hammer to form the first corner. Hammer the wire against the side of the vise to complete the right angle.

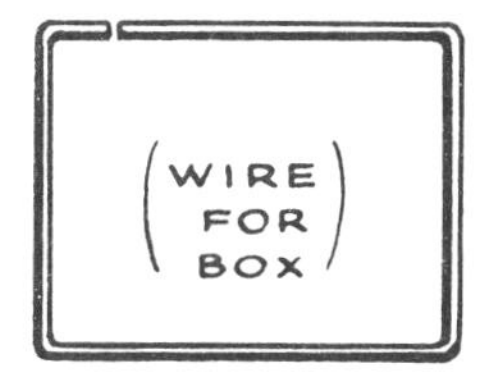

Figure 33-15
Wire for box edge.

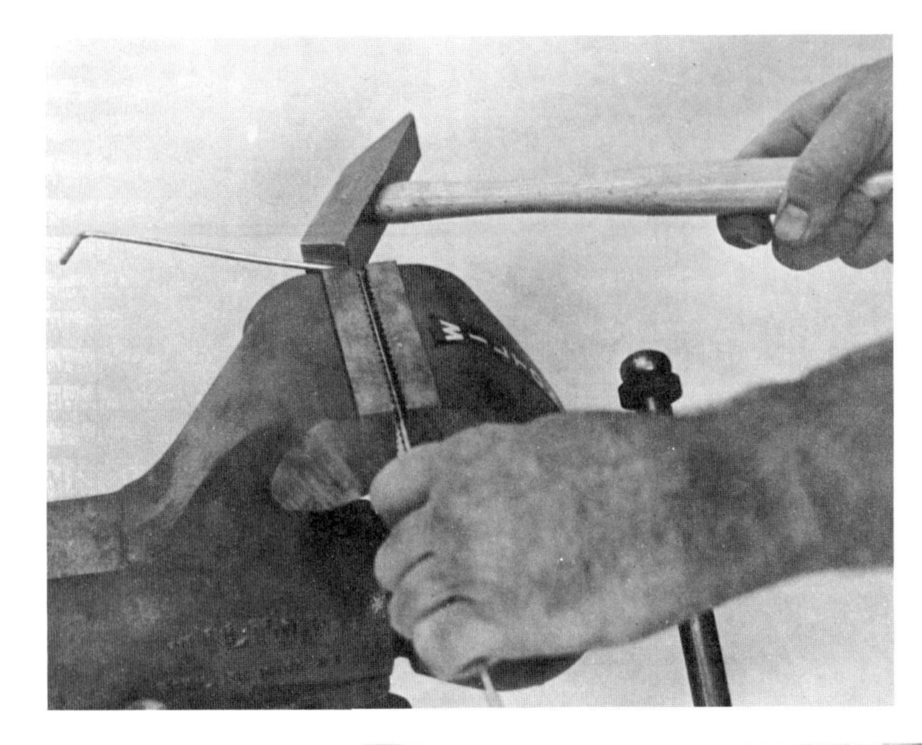

Figure 33-16
Bending the wire in a vise.

Figure 33-17
Starting the fold.

5. Reposition the bent part in the vise and bend the wire to form the second corner, figure 33-16.
6. Form the third and fourth corners of the frame by repeating step 5.
7. Place the wire frame on the box.
8. Lay the job on the square end of the hollow mandrel stake or any other suitable stake.
9. Start the fold over the wire at each corner, using the mallet and pliers, figure 33-17.

Use pliers to prevent injury to your fingers.

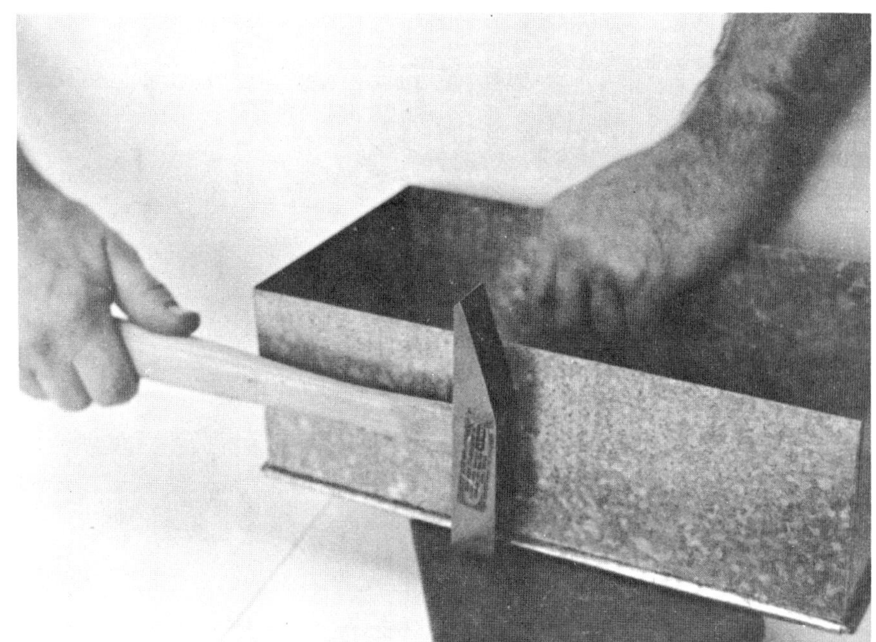

Figure 33-18
Bending metal over wire.

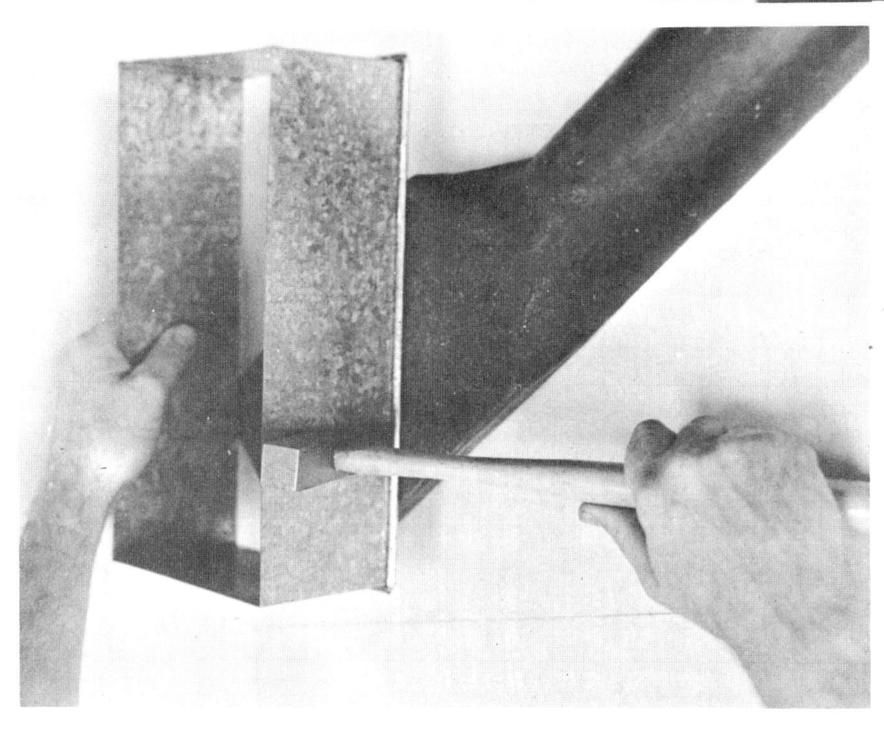

Figure 33-19
Completing the wired edge.

10. Bend the metal over the wire as far as possible with the mallet on each of the four sides.
11. Bend the metal over the wire as far as possible with the face of the setting hammer, figure 33-18.
12. Complete the wired edge by peening the metal down tightly over the wire on each of the four sides with the peen of the setting hammer, figure 33-19.

Be careful not to mar the job.
Solder the corners and file them round and smooth.

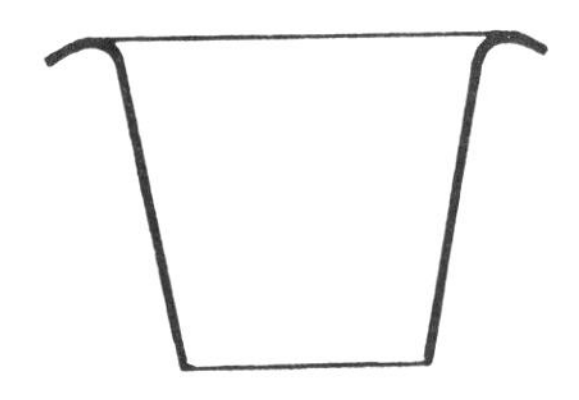

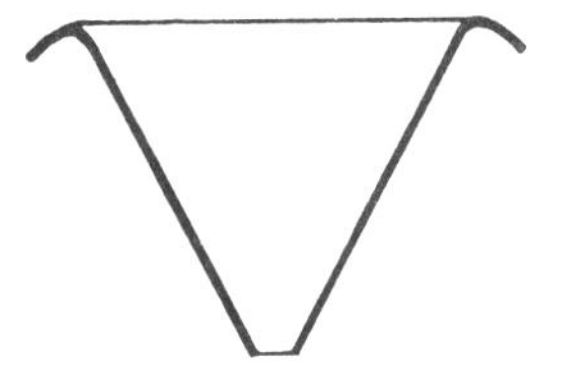

Figure 33-20 Forming taper edge for wire.

HOW TO WIRE A TAPERED JOB AFTER IT IS FORMED

1. Turn the edge of the job equal to 2-1/2 times the diameter of the wire on the turning machine. Figure 33-20 shows an edge turned on a slightly tapered and on a sharply tapered job.
2. Cut the wire to the correct length with a pair of pliers or wire nippers.

Wire should not be cut with snips or on the squaring shear because it nicks the cutting edge.

3. Form the wire into a circular shape in the groove on the slip roll forming machine.
4. Place the job on the hollow mandrel stake.

CAUTION

When wiring a funnel, use the blowhorn stake.

5. Start turning the metal over the wire with a mallet and a pair of pliers, figure 33-21.

Start the wire 1 to 2 inches from the seam. This is done to strengthen the seam and reduce the possibility of an opening remaining after the job is formed.

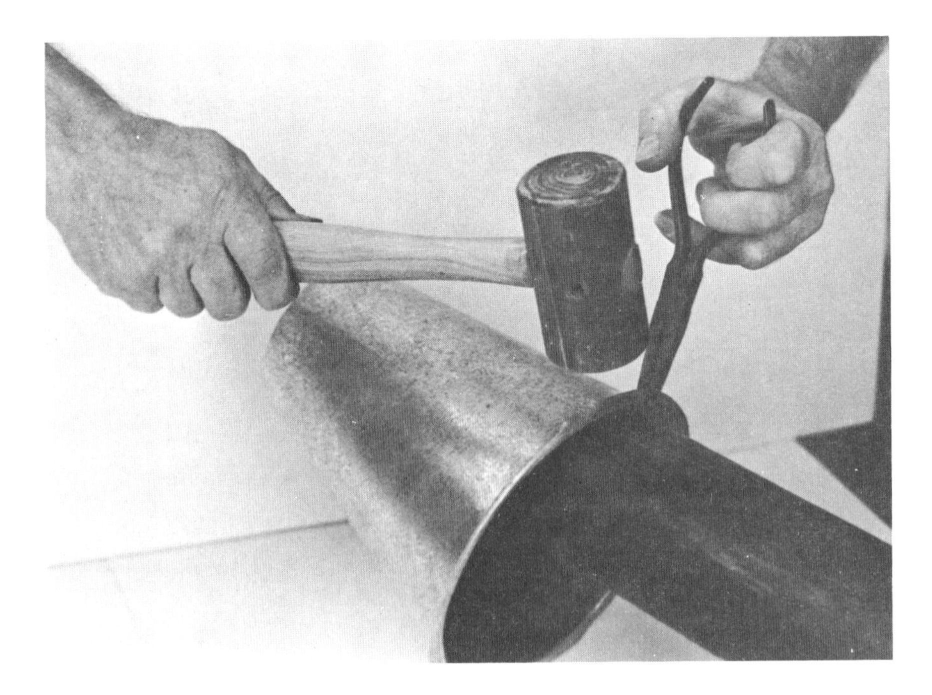

Figure 33-21 Starting the wire edge on a tapered job.

Figure 33-22 Bending the metal over wire on a tapered job.

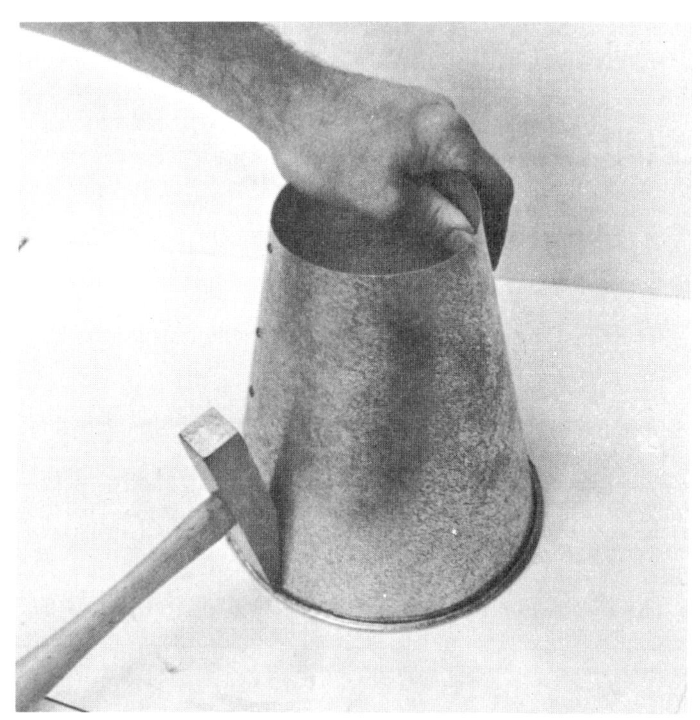

Figure 33-23
Finishing the peen.

6. Bend the metal over as far as possible with the flat face of the setting hammer, figure 33-22.
7. With the peen of the setting hammer, complete the wired edge by peening the metal down tightly over the wire, figure 33-23.

Be careful not to mar the job with the mallet or the setting hammer.

SUMMARY REVIEW

A. Place the answers to the following questions in the column to the right.

1. Give three reasons for putting a wired edge on a job. 1. ______________________

   ____

   ____

2. List two ways wire edges can be made. 2. ______________________

   ____

3. List two jobs that should be wired before they are formed. 3. ______________________

   ____

4. Give three examples of jobs wired after they are formed. 4. ______________________

   ____

   ____

5. State two methods of ordering wire from the mill. 5. ______________________

   ____

B. Insert the correct word in each of the following sentences.

1. Wire obtained in straight lengths is called ______________ .
2. The wire edge is always put on the ______________ of a job unless otherwise specified.
3. The edge allowance on jobs of 24 gage and lighter is ______________ times the ______________ of the wire.
4. The formula for the wire length for the edge of a square job can be expressed as ______________ .
5. A wire edge can be prepared to receive the wire by forming in the ______________ or ______________ .

C. Underline the correct word in each of the following sentences.

1. Wire should be cut with (snips, squaring shear, wire nippers).
2. The beginning fold of the metal over the wire is done with pliers and a (setting hammer, riveting hammer, mallet).
3. Always place the wire to (lap, fold, extend) 1 to 2 inches from the underfold end.
4. Pliers are used to hold the wire to the metal to prevent (slipping, marring, bending, injury).
5. In making a wire edge on an 8-inch square pan of 26-gage galvanized iron with 1/8-inch diameter wire, a length of (32, 32-1/8, 32-1/4) inches is used for the wire.

IDENTIFICATION REVIEW

Directions: Identify each tool and print its name below or along side of the picture of the tool.

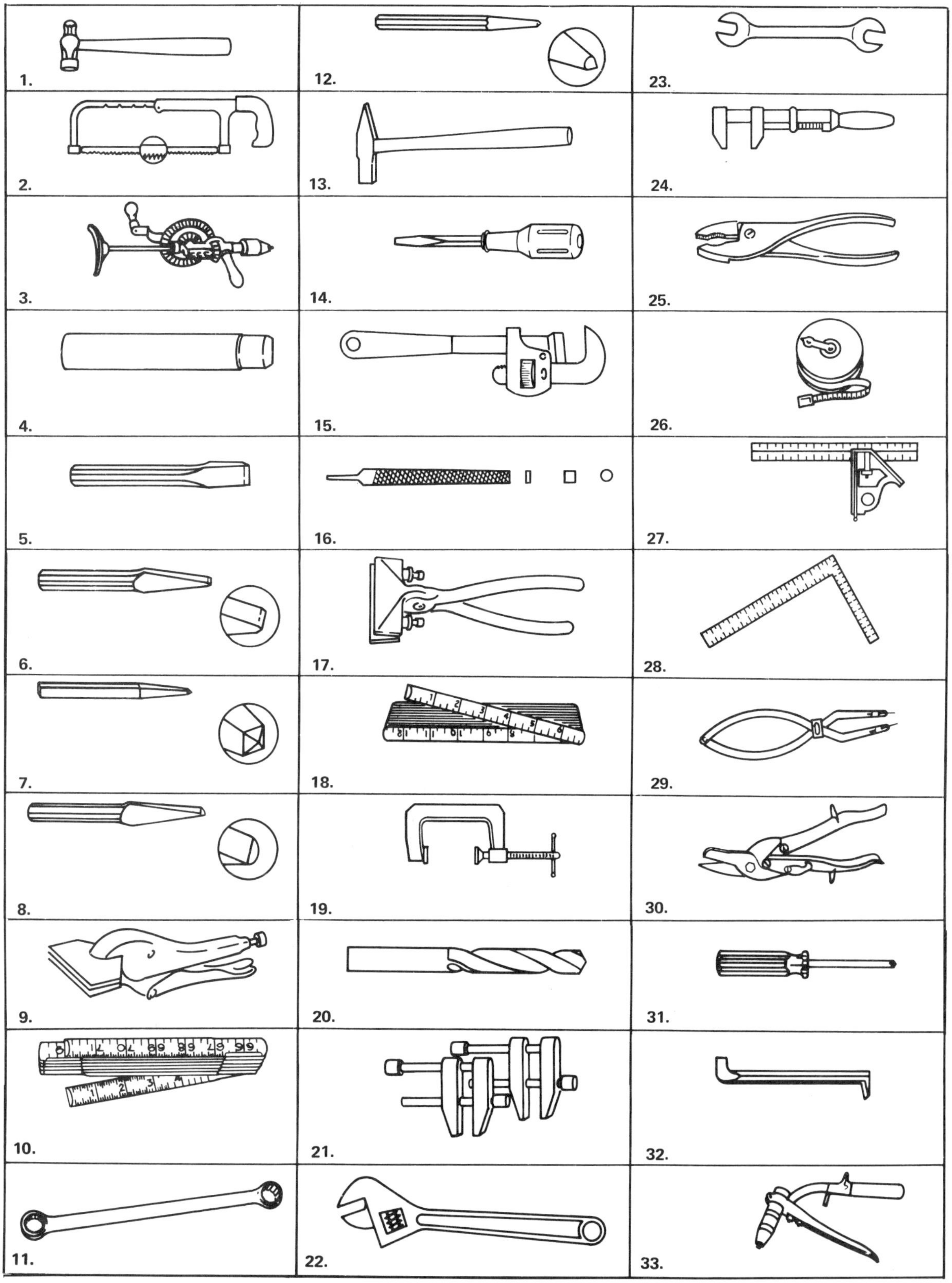

1.	12.	23.
2.	13.	24.
3.	14.	25.
4.	15.	26.
5.	16.	27.
6.	17.	28.
7.	18.	29.
8.	19.	30.
9.	20.	31.
10.	21.	32.
11.	22.	33.

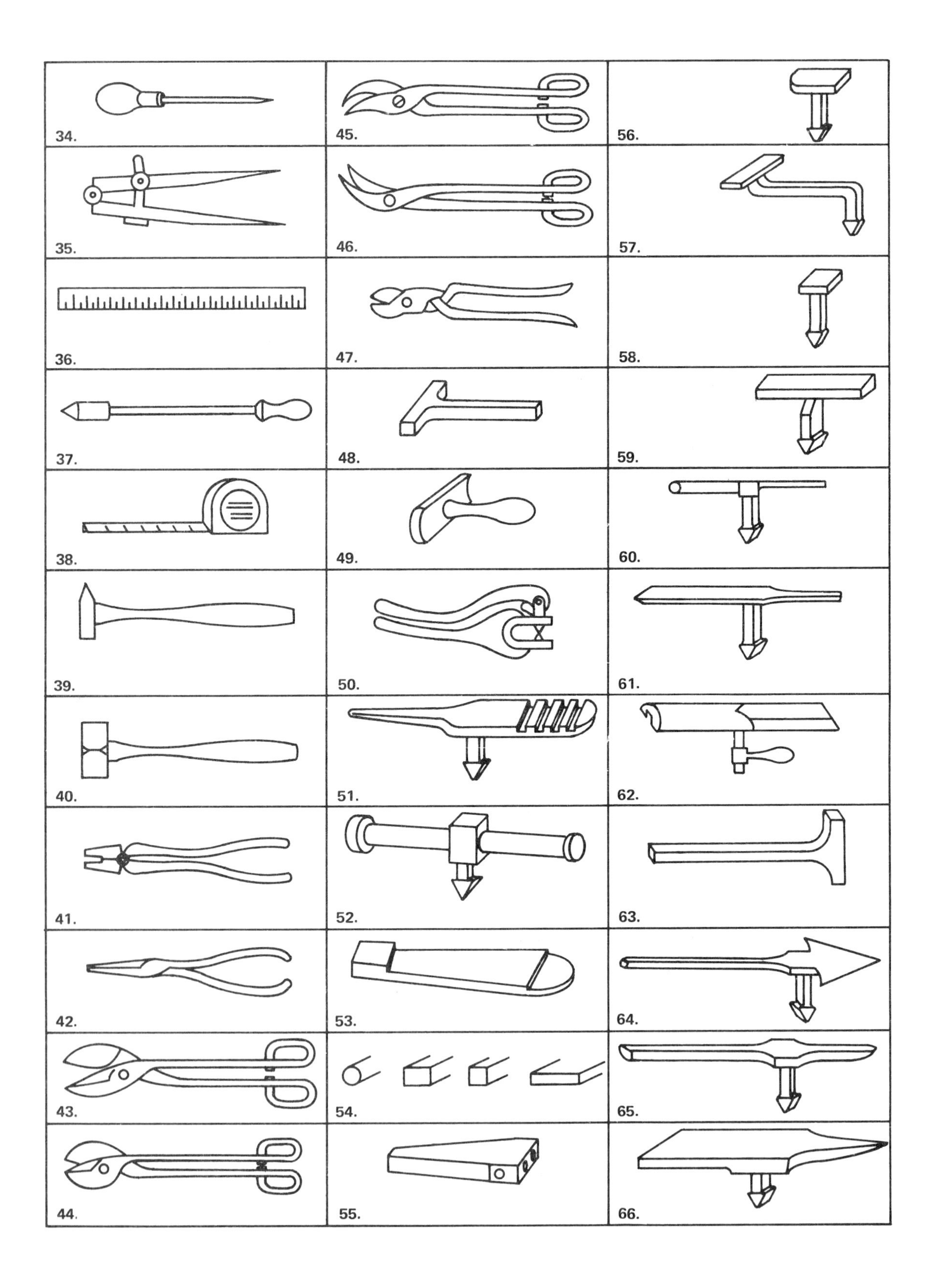
34.
35.
36.
37.
38.
39.
40.
41.
42.
43.
44.
45.
46.
47.
48.
49.
50.
51.
52.
53.
54.
55.
56.
57.
58.
59.
60.
61.
62.
63.
64.
65.
66.

DIRECTIONS:

a) Identify each seam or edge.

b) State where used as flat, corner, round, raw edge, bottom.

c) Give total allowance formula using W for seam width, d for rivet or wire diameter.

67	a. ______ b. ______ c. ______	73	a. ______ b. ______ c. ______
68	a. ______ b. ______ c. ______	74	a. ______ b. ______ c. ______
69	a. ______ b. ______ c. ______	75	a. ______ b. ______ c. ______
70	a. ______ b. ______ c. ______	76	a. ______ b. ______ c. ______
71	a. ______ b. ______ c. ______	77	a. ______ b. ______ c. ______
72	a. ______ b. ______ c. ______	78	a. ______ b. ______ c. ______

DECIMAL EQUIVALENT TABLE

Fraction	Decimal	Fraction	Decimal
1/64	.015625	33/64	.515625
1/32	.03125	17/32	.53125
3/64	.046875	35/64	.546875
1/16	.0625	9/16	.5625
5/64	.078125	37/64	.578125
3/32	.09375	19/32	.59375
7/64	.109375	39/64	.609375
1/8	.1250	5/8	.6250
9/64	.140625	41/64	.640625
5/32	.15625	21/32	.65625
11/64	.171875	43/64	.671875
3/16	.1875	11/16	.6875
13/64	.203125	45/64	.703125
7/32	.21875	23/32	.71875
15/64	.234375	47/64	.734375
1/4	.2500	3/4	.7500
17/64	.265625	49/64	.765625
9/32	.28125	25/32	.78125
19/64	.296875	51/64	.796875
5/16	.3125	13/16	.8125
21/64	.328125	53/64	.828125
11/32	.34375	27/32	.84375
23/64	.359375	55/64	.859375
3/8	.3750	7/8	.8750
25/64	.390625	57/64	.890625
13/32	.40625	29/32	.90625
27/64	.421875	59/64	.921875
7/16	.4375	15/16	.9375
29/64	.453125	61/64	.953125
15/32	.46875	31/32	.96875
31/64	.484375	63/64	.984375
1/2	.5000	1	1.0000

INDEX

7/90(9C1702I)